THE SHAPE OF SPACE

DINING SPACES

THE SHAPE OF SPACE

DINING SPACES

JAMES DARTFORD

VNR SPRINGER SCIENCE+BUSINESS MEDIA, LLC

Acknowledgements

I should like to express my thanks to:

Jeffrey Serlin, of N.A.S. Furniture Contracts Ltd., London, England for some most useful tips on the custom furniture industry.
Martin Morris, of Primo Furniture Ltd, Enfield, England for initiation into table/seat combination units and the self-service market.

Christopher and Elly King of King and King Design Associates: for tracing, annotating and upgrading my original drawings with consummate skill and attention to detail.
Bruce Robertson, of Diagram Visual Information Ltd: for thinking up the series in the first instance, for keeping my nose to the grindstone ever since, and for putting the book together.

Originally published by Van Nostrand Reinhold in 1990
Softcover reprint of the hardcover 1st edition 1990

Library of Congress Catalog Card information is available.
ISBN 978-1-4757-4996-0 ISBN 978-1-4757-4994-6 (eBook)
DOI 10.1007/978-1-4757-4994-6

16 15 14 13 12 11 10 9 8 7 6 5 4 3 2 1

INTRODUCTION

This book is primarily concerned with the space planning of restaurants and, since their basic elements are much the same throughout the civilised world, is directed at an international readership.

Although concentrating on the average restaurants, we also look at the wider commercial range – banquet rooms, cafeterias, fast food establishments, drug stores, sandwich bars and cafes – as well as private dining rooms, outdoor and in-transit situations. Where a particular environment is not specifically singled out, its elements can be deduced by comparison to similar spaces.

At a detailed planning level, the design of dining spaces is not a well investigated and documented field. The present work attempts to fill this gap and address the designer's need for hard facts, dimensions, and sound spatial guidelines.

The author of this book firmly believes that given rigorous scrutiny of all relevant data almost any design problem can be solved; and that the process of arriving at a solution (or recommendation) can and should be explicable in unequivocal terms.

It is hoped as a result that this book will be useful to restaurateurs and foodservice specialists as well as to architects and interior designers.

GUIDE TO USING "DINING SPACES"

Various aspects of the book's structure, illustrations and text may usefully be clarified:

1 **Information retrieval**: each Section opens with a subject breakdown of the CONTENTS. These detailed sub-divisions, accompanied by an identifying pictogram, form the titles of each illustrated sheet and obviate the need for a separate index.

2 **Continuity**: the end-of-Section SUMMARIES are intended to reinforce the main thread of investigative argument and to introduce the ensuing topic.

3 The fact of addressing an international audience has necessitated taking certain linguistic and dimensional decisions that inevitably favor one side of the Atlantic or the other:

3.1 **Language:** spelling conforms to AMERICAN English and, with possibly one or two exceptions, words have been selected that should be equally understood by all English-speaking peoples.

3.2 **Dimensions:** CENTIMETERS are primarily adopted throughout, because the metric system is far more widespread and represents an improvement over American "customary units" and former British "imperial." In the context of the book, centimeters are clearly preferable to millimeters.

These metric dimensions are, however, accompanied (in *italics*) by their RATIONALIZED EQUIVALENT in inches – or, above 8ft, in feet and inches. This approximate conversion, based on **10cm** = **4"** differs from the strictly correct translation of 10cm = $3\tfrac{15}{16}$, or 4" = 10.16cm, by a negligible 1.6%; and the reasons for adopting it should be obvious.

All nations think, and therefore manufacture, in round figures. A foreign table 120cm long equates with an American one of 4'0" length (not 3'$11\tfrac{5}{8}$"), and a table of 30" diameter will find its European equivalent at 75cm (not 76.2cm).

Moreover, in a space planning manual such as this, no dimension whatsoever merits the irritating exactitude of a micrometer.

3.3 **Scale:** following suit, all drawings are at scales of 1:20, 50, or 100. (These approximate quite closely, of course, to $\tfrac{1}{2}$" (1:24), $\tfrac{1}{4}$" (1:48) and $\tfrac{1}{8}$" (1:96) to 1ft.)

4 **Symbols:** to keep the graphics as simple as possible, and to avoid undue repetition, the normal architectural drawing conventions have been supplemented by certain symbols or abbreviations. Examples, with attached notes, are shown below:

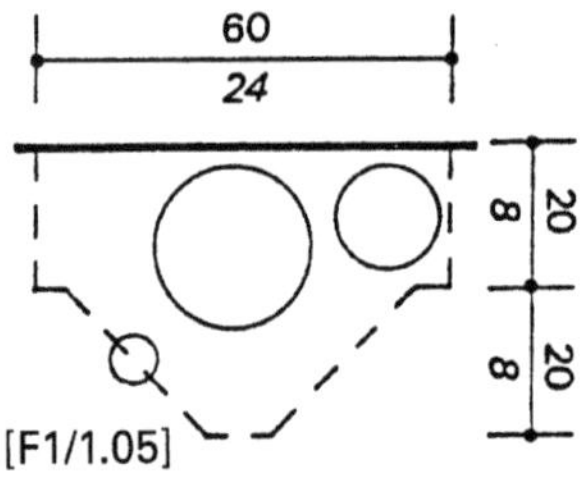

[F1/1.05]

[F/1.02]

Place settings: three items – dinner plate, side plate and glass – imply, in graphic shorthand, associated cutlery and perhaps a further glass. Dimensions, shown in the first figure, are typical – centimeters and (in *italics*) rationalized equivalent inches.

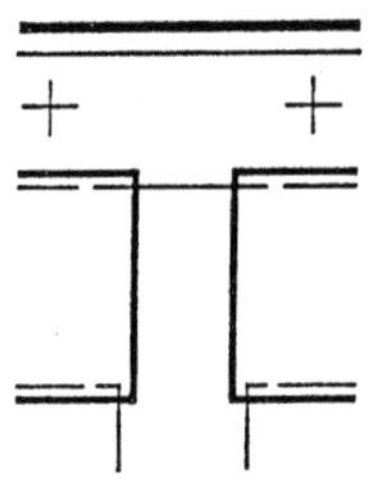

[D/3.02]

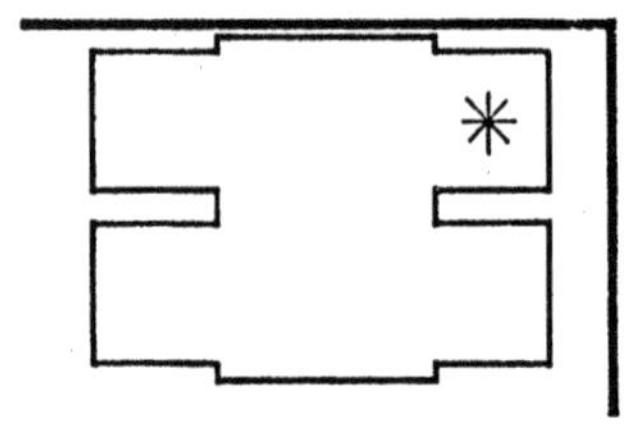

[G1/5.10]

Left: crosses denote individual bench seat locations, and the bold central figure (**4**) represents the number of seats per table. The number of persons accommodated in larger configurations are similarly indicated.

Right: seats asterisked signal special mention in the accompanying text. (In this case, an access problem.)

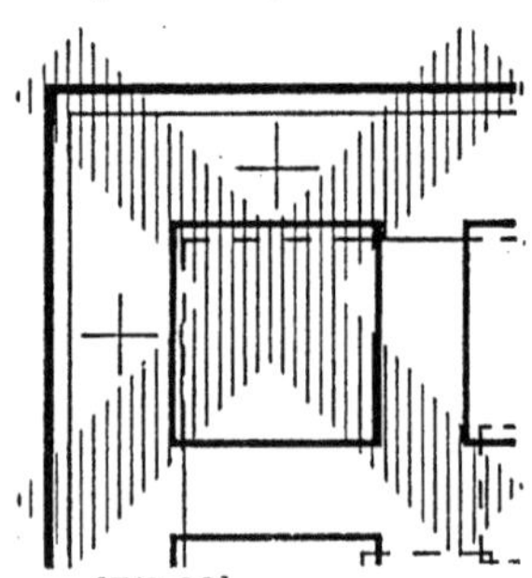

[F/3.02]

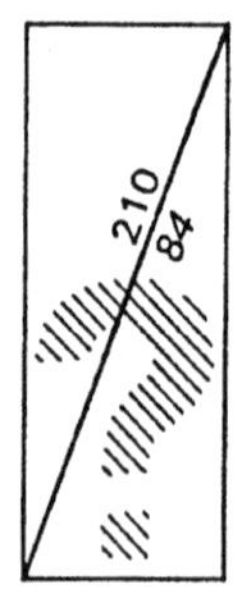

[K2/2.04]

Left: design solutions NOT recommended are overlaid with a cross. (In this case, an unaccessible corner table.)

Right: overlaying with a question mark raises a query. (In this instance, whether a particular table top will pass a standard door.)

CONTENTS

DINING COMFORTABLY

1

DINING COMFORTABLY
Anthropometric criteria: person at table

1.01

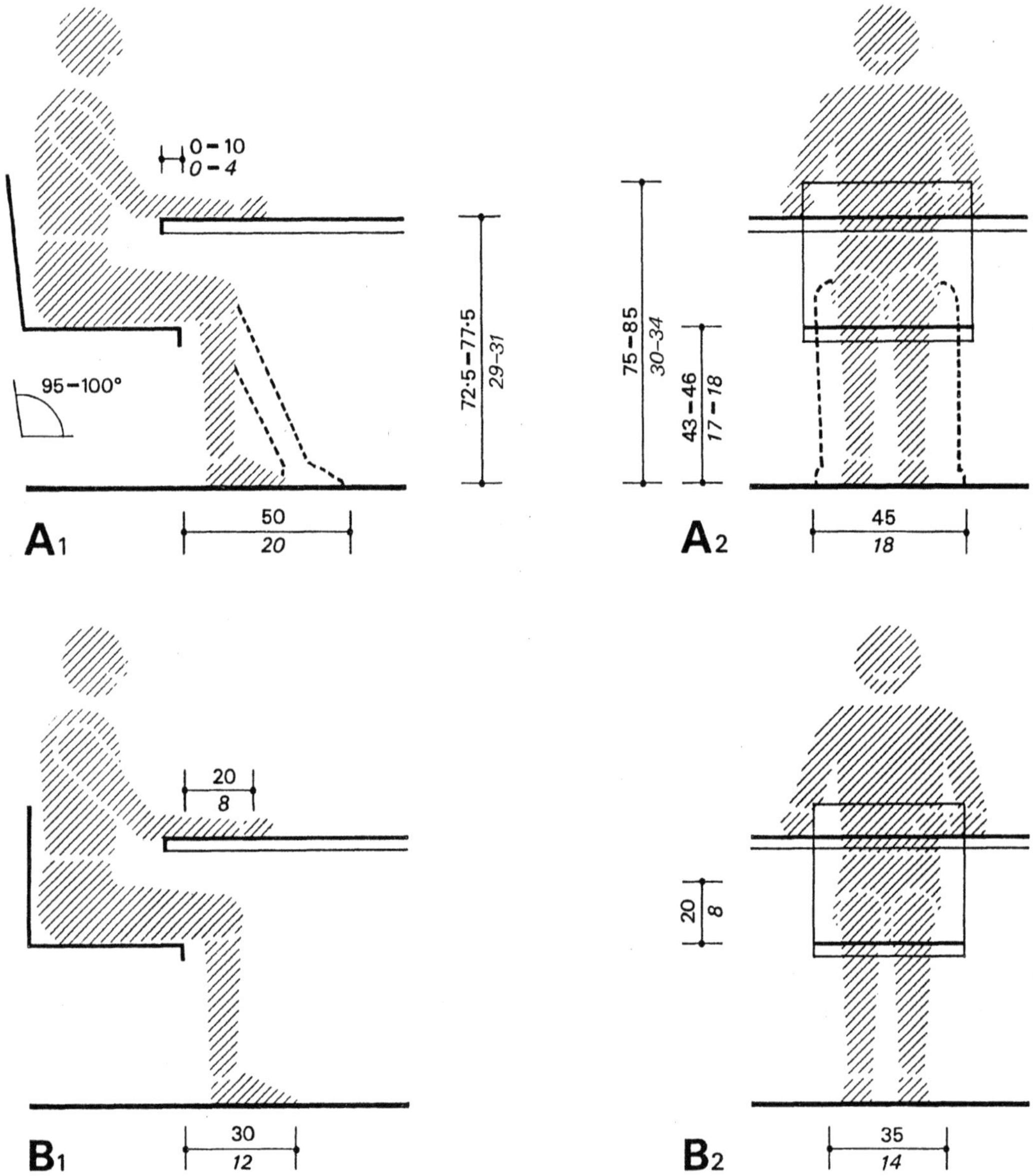

1:20

- Sitting comfortably at a dining table is an interrelated compound of correct posture, furniture shape and dimension.
- Over the centuries, the heights of upright chairs and tables have scarcely fluctuated, and a common rule of thumb is to allow 45cm/*18"* and 75cm/*30"* respectively.
- Other widely accepted measurements, shown here, concern seat widths and back slope, knee room, leg room and table – seat overlap.

A1, A2 Generous seating, with plenty of forward and lateral leg room and sufficient personal tableware depth.

B1, B2 Adequate seating, with minimum leg room. Note that thigh clearance becomes an important consideration with deep-edged refectory tables.

Note: The human figures, in both upper and lower views, represent an average male adult. Dimensional allowances may be expected to cater for nine out of ten persons.

DINING COMFORTABLY

Anthropometric criteria: primary design variables

1.02

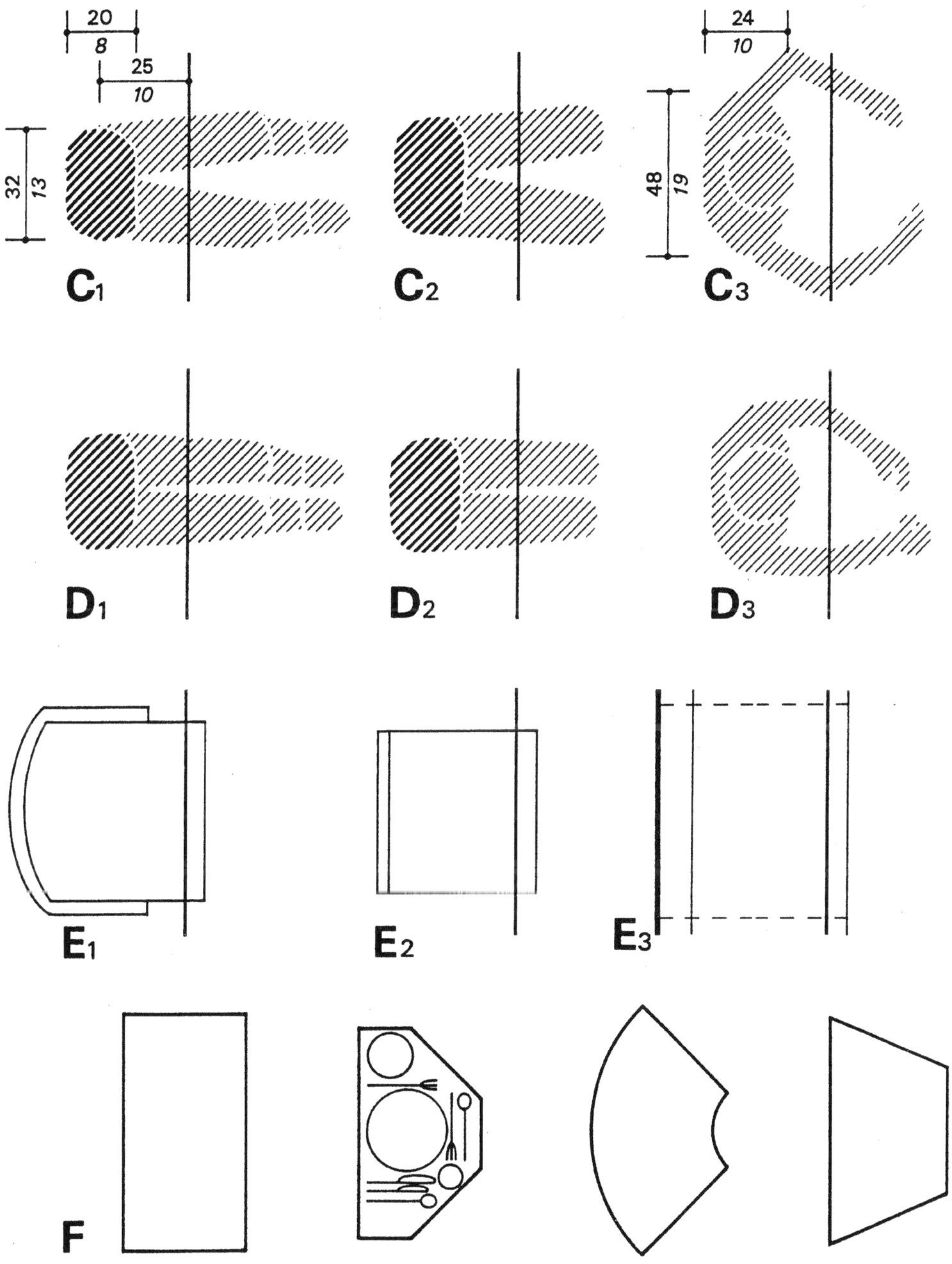

1:20

- Leg and feet positions, elbow room, extent of reach, seat size and a range of place setting allowances are illustrated. Any one of these variables, or a combination of them, will serve to test the suitability of a provisionally-selected table for the targeted type of food service. These are major determinants in the table-sizing exercises that follow in the next Section.

C1 Below-waist body dimensions, knees apart and legs outstretched.
C2 Knees apart, legs retracted.
C3 Above-waist body dimensions, wide elbow room.
D1, D2 Knees together, legs progressively retracted.
D3 Restricted elbow room.
E1 Generous armed chair.
E2 Minimal armless chair.
E3 Personal area of bench seat.
F Personal place setting allowances as dictated by size and shape of table.

G1

G2

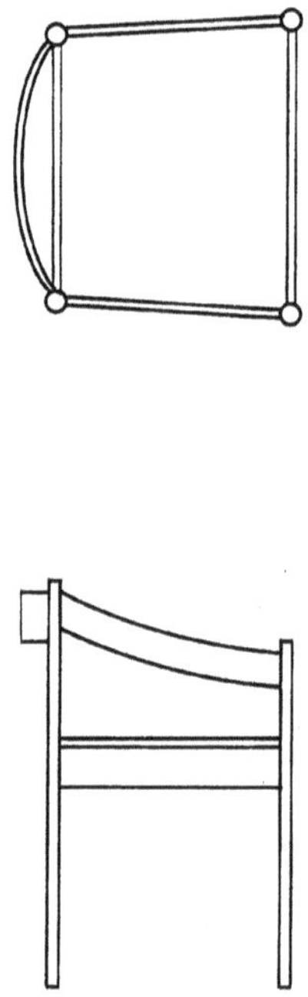

G3

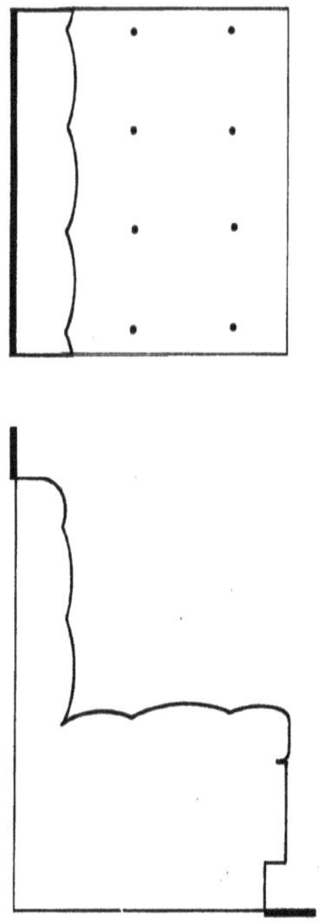

H1

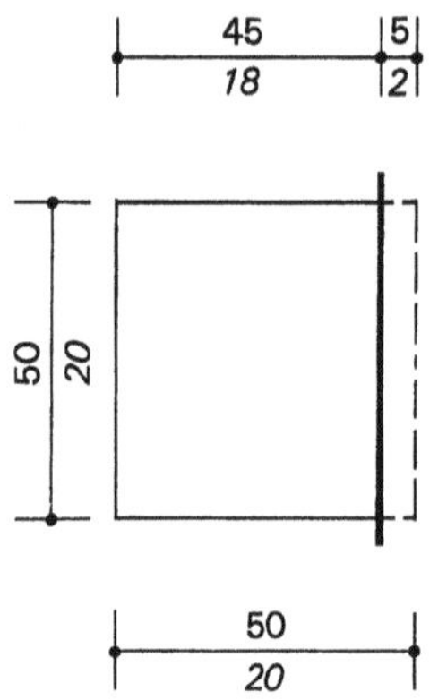

G4

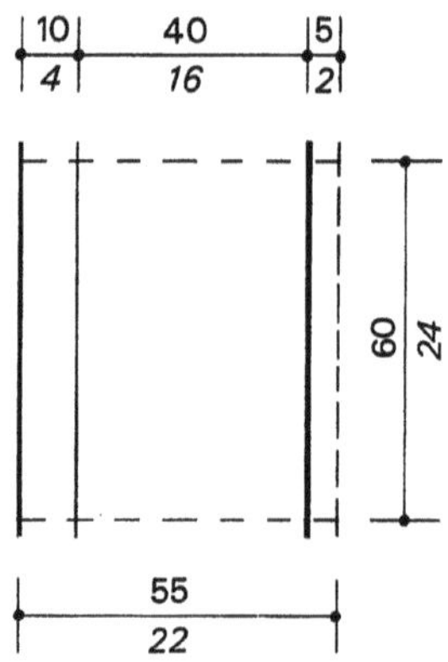

H2

1:20

- In the context of dining tables we are concerned with two types of seat: upright chair and bench. (Stools, in conjunction with counters, are introduced in Section 4.)
- Chairs, despite an identical function, vary considerably in their shape and structure.
- Benches, on the other hand, differ mainly in their degree of comfort: ranging from plain and hard to upholstered and soft.
- For the purposes of this book, certain average seat dimensions, including table/seat overlap, may be assumed as design tools.

G1 Anonymous straight-backed chair.
G2 Marcel Breuer's classic Bauhaus tubular chair.
G3 Magistretti's equally classic beech armed chair.
G4 Chair: "Design tool" dimensions, in plan.
H1 Upholstered bench seat (banquette).
H2 Bench: "Design tool" dimensions, in plan.

DINING COMFORTABLY
Place settings: introduction.

1.04

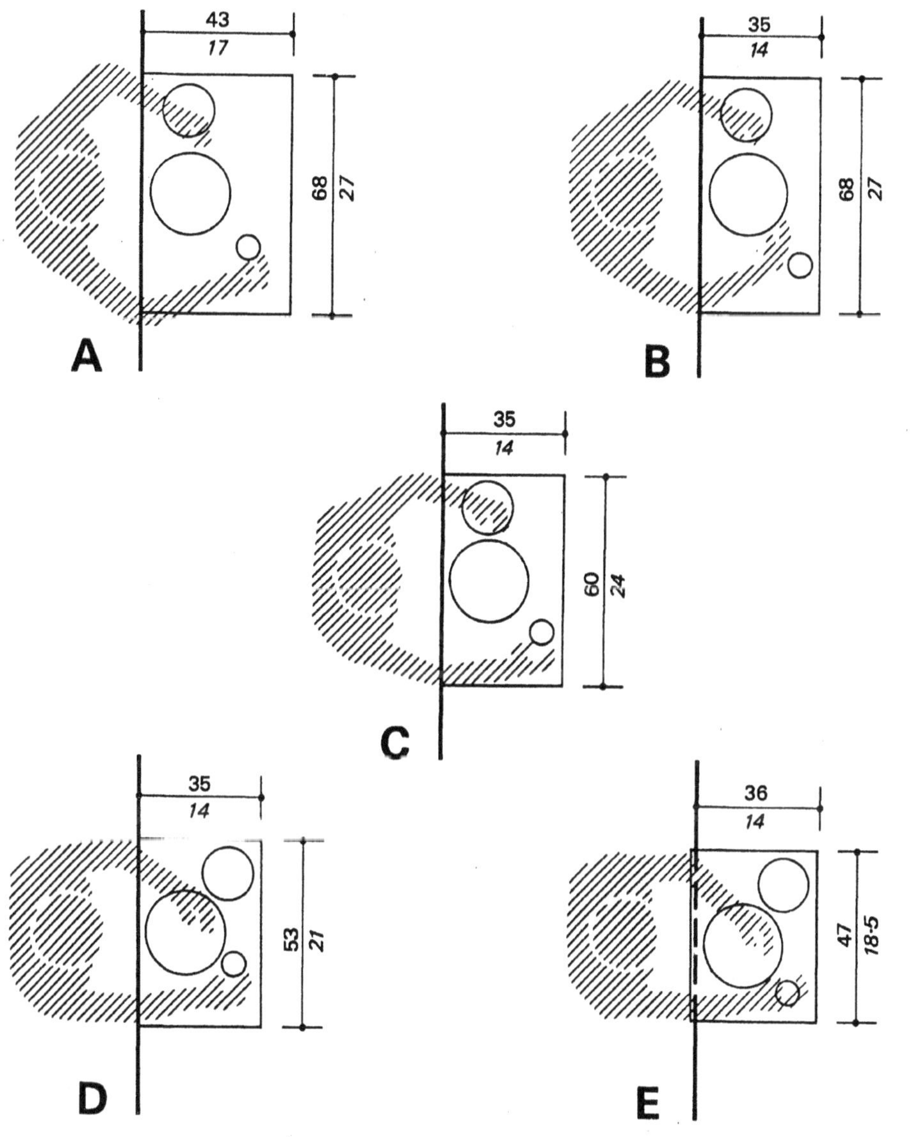

1:20

A range of typical rectangular place settings, shown with an individual's basic tableware (cutlery excluded).

A As recommended by an authoritative anthropometrist, but commercially impracticable.
B Ideal place settings.
C Accepted commercial norm.
D Place setting common in self-service establishments.
E Standard lipped cafeteria tray.

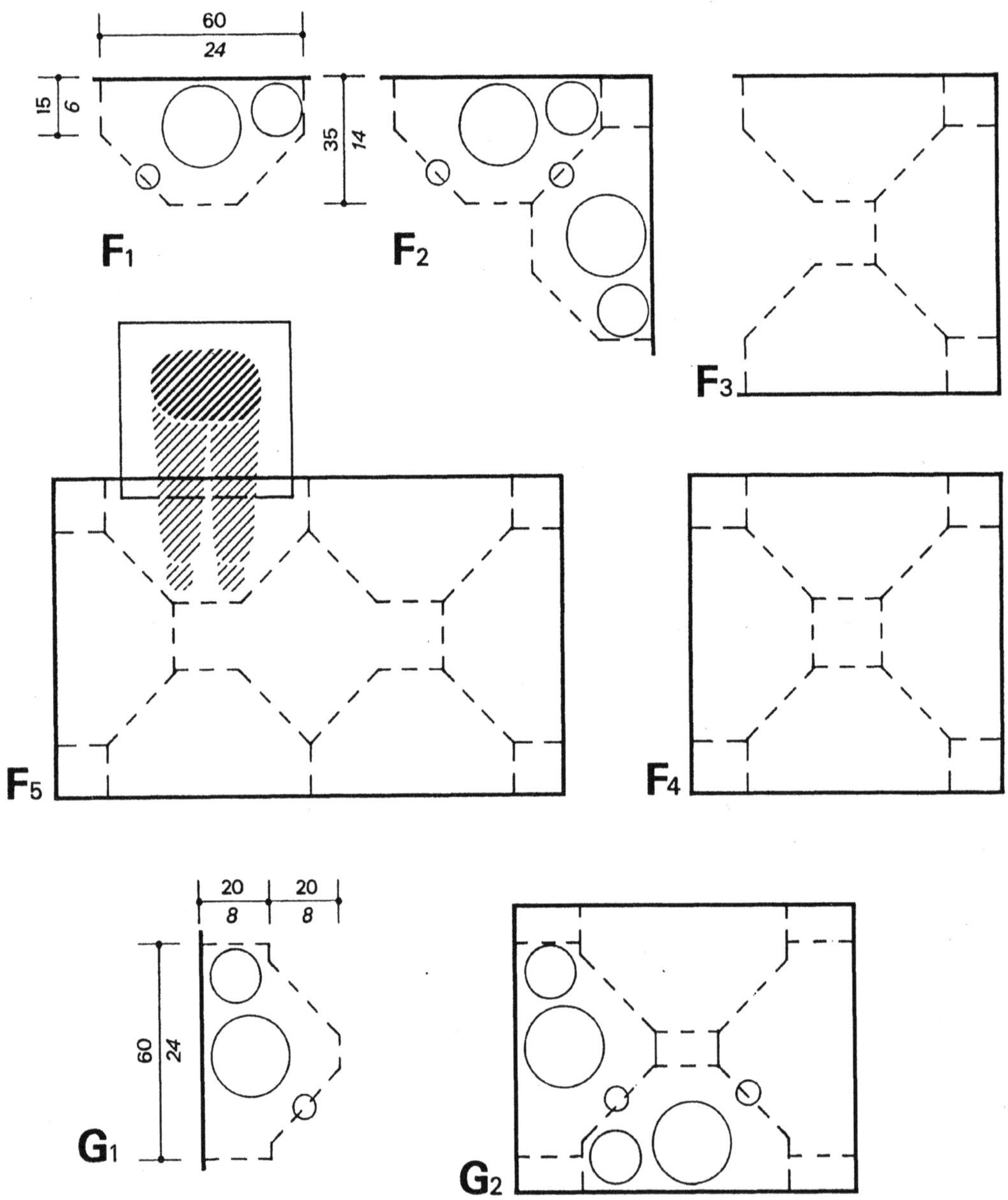

1:20

Space is at a premium at the ends of tables or around a square. Place settings (considerably larger than table mats) may however be cut-cornered with no real detriment. This notional device establishes "extra" length for rectangular tables and correct dimensions for square ones.
Note: the residual communal surfaces, in the middle of the table.

F1 Chamfered shape, determines:
F2–F5 table sizes for 2, 3 and up to 6 persons.
G1, G2 Variant end setting for narrower tables.

Place settings: round, hexagonal and octagonal tables.

1.06

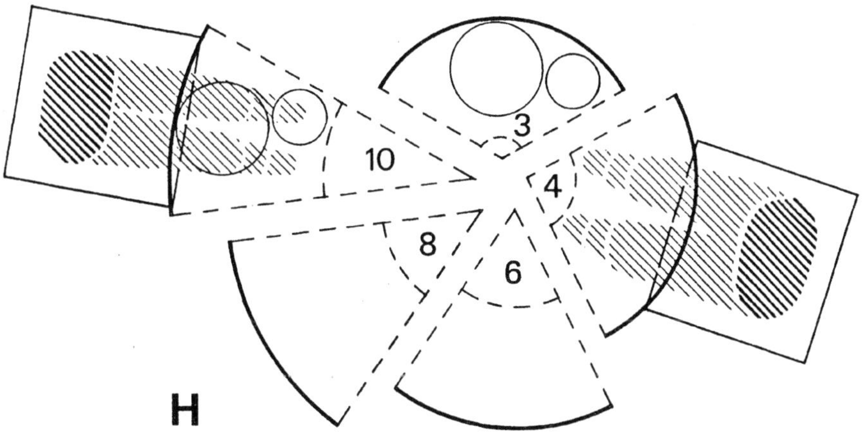

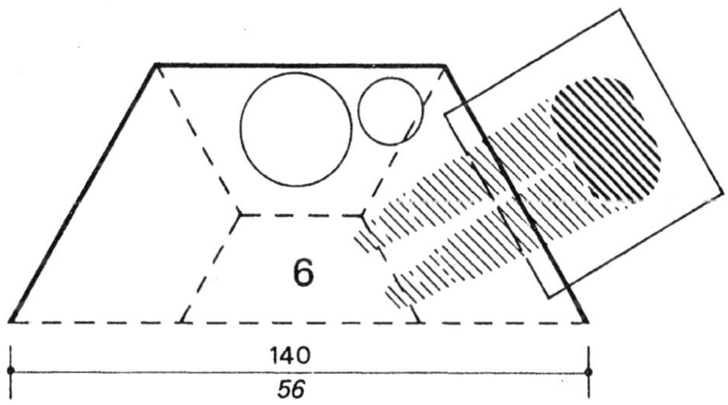

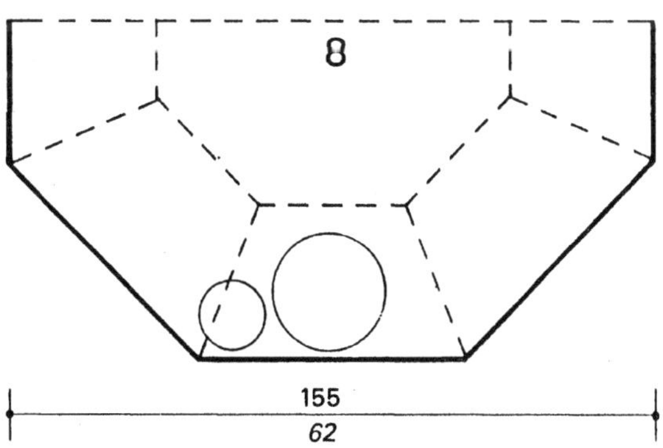

1:20

- Individual "places" must undergo further changes of shape with round, hexagonal or octagonal tables.
- Whereas persons seated at appropriately-sized hexagons and octagons have ample leg room, those at round tables will be slightly more constricted. A further corollary of the pie shape is that place width diminishes with increased table size. This poses no great problem, unless to a purist: with a place setting increased to say 45cm/*18"*, an individual's side plate simply moves forward towards the middle.
- Since dinner plates are centrally positioned, the extremity of one person's right-hand space may be borrowed by his right-hand neighbor.

H A composite plan, representing round tables for 3, 4, 6, 8 and 10 persons, with respective diameters of 80cm/*32"*, 100cm/*40"*, 125cm/*50"*, 150cm/*60"* and 180cm/*72"*.

I Composite plan, representing a suitable hexagonal table for 6 persons.

J Composite plan, representing a suitable octagonal table for 8 persons.

DINING COMFORTABLY: summary

1. Dining comfortably at a table is a matter of degree; and the degree of comfort depends on pre-planned circumstances.
2. Table and seat heights, largely established by tradition, are virtual constants.
3. Posture, legroom and kneeroom are the anthropometric variables, for which certain minimum dimensions are relevant.
4. Table shape and size, seat distribution, and type of meal are further functional variables that, together with elbowroom, influence an individual's place setting or "cover."
5. Regarding the type of meal, one may distinguish between two extremes: renewability (of food dishes and tableware, in restaurant service) and once-and-for-all provision (in cafeterias and fast food outlets).
6. A reasonable average place setting may be contained within a notional 60×35cm/*24″×14″* rectangle, but this shape must be adapted to turn corners or to fit round, hexagonal or octagonal tables.

The next Section applies these findings to a wide variety of separate, modular, linked and convertible tables, and to combined table/seat units.

TABLES AND CHAIRS

2

TABLES AND CHAIRS
Separate tables: rectangular, side-seating only

2.01

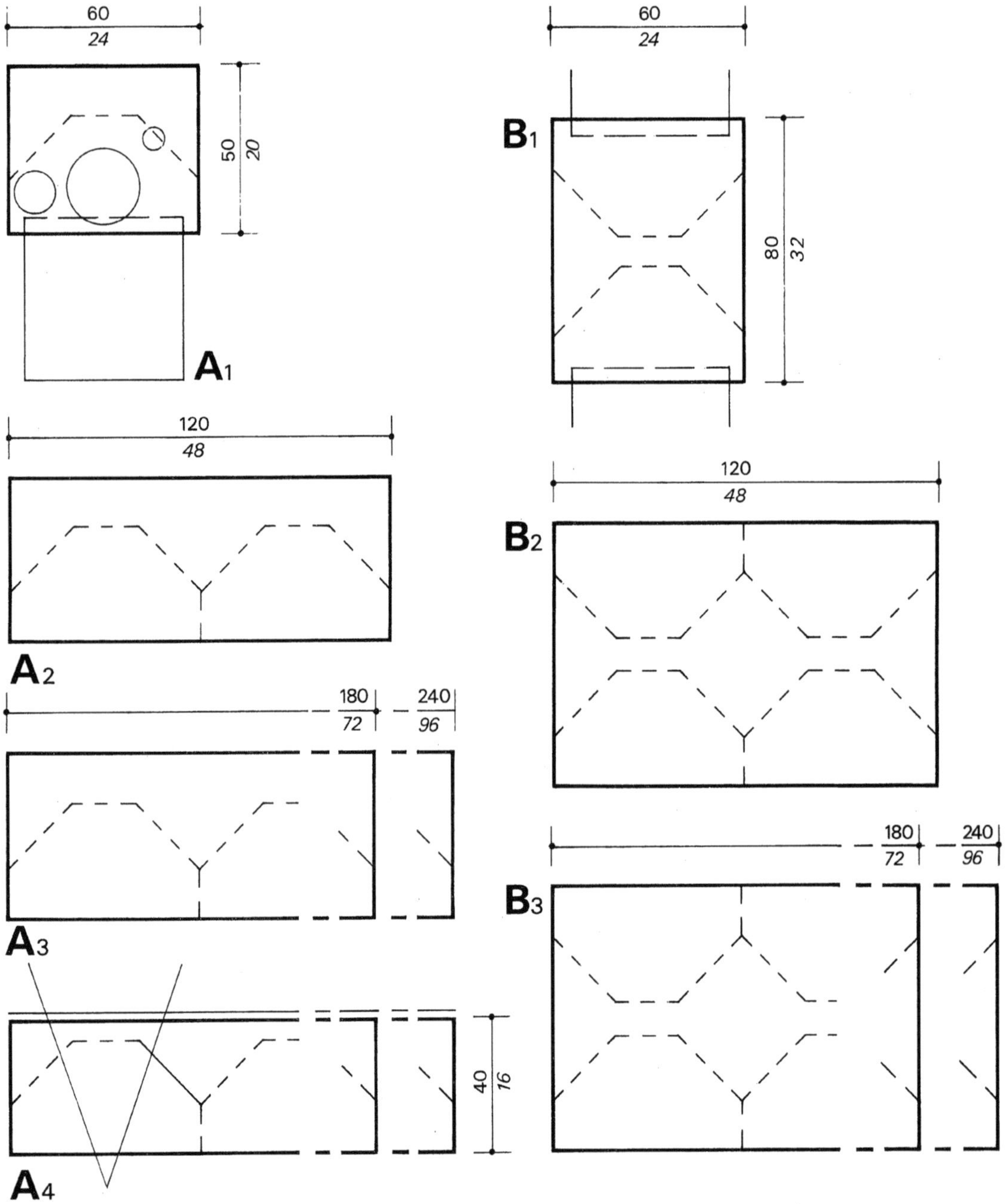

1:20

Note: These – and successive figures – are based on considerations examined and dimensions established, in the preceding section.

A1–A3 Unilaterally-seating tables. Suitable only where space is limited, or facing a wall. To increase depth beyond 50cm/*20″* would encourage opposing seats, with resulting knee-clash and fouling of access aisle.

A4 An even narrower built-in table surface, perhaps behind glass, will serve in "spectator dining" situations such as racing and Ice Follies.

B1–B3 Bilaterally-seated tables. A table depth of 80cm/*32″* provides reasonable foot room, and a fair balance between individual and shared table surface. It is worth noting that a basic module of 60cm/*24″* and 80cm/*32″* may be taken, without wastage, from universally standard 240cm × 120cm/*8′ x 4′* boards.

2.02

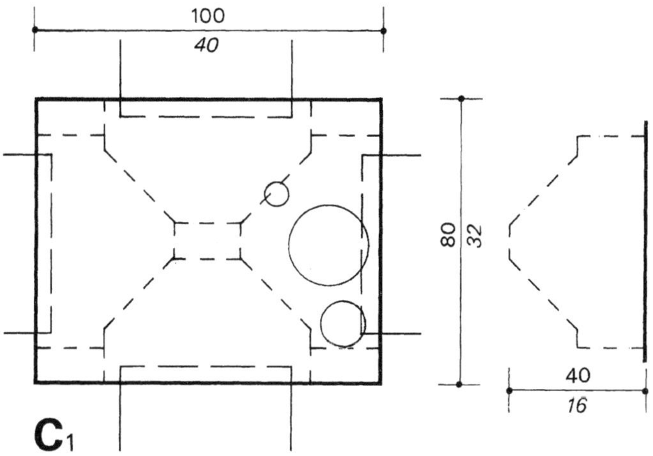

C1

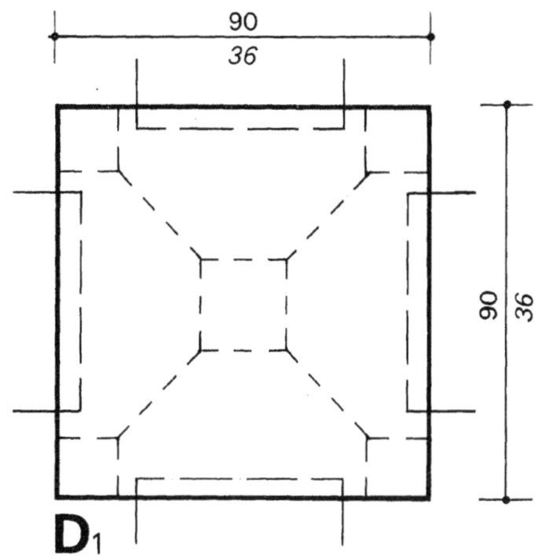

D1

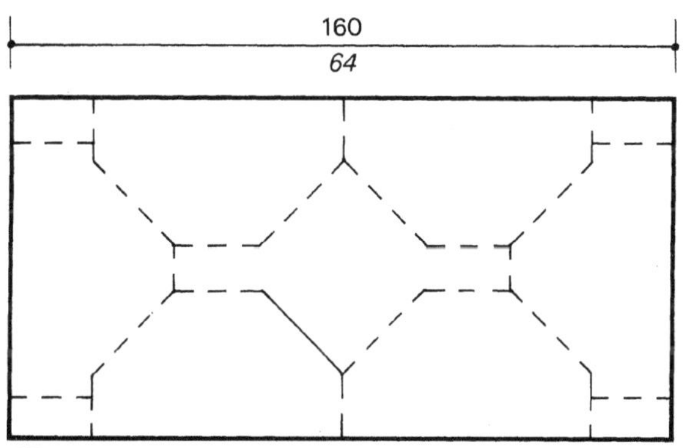

C2

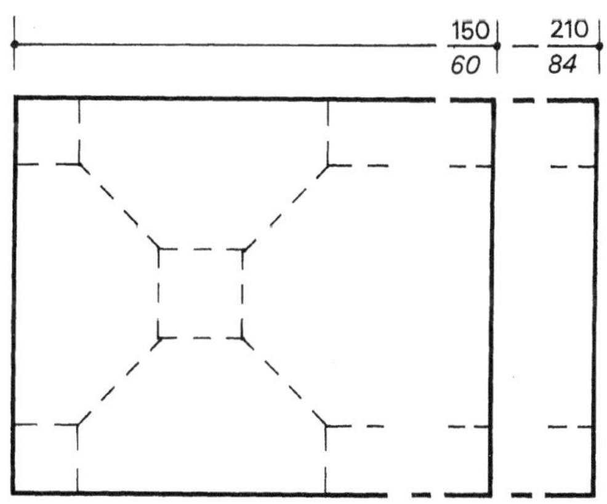

D2

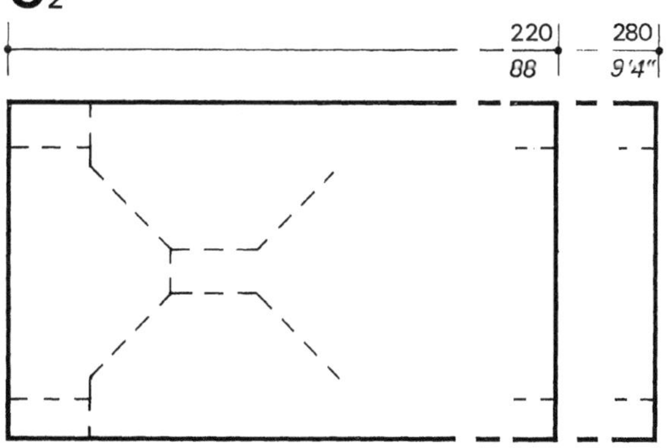

C3

1:20

C1–C3 Side- and end-seat tables 80cm/*32"* wide. When the table is for 4 persons only, there is little central space. Otherwise, an efficient and economic range. Note the end place variant.

D1, D2 Side- and end-seat tables 90cm/*36"* wide, that compare favorably with the previous examples and include the common 4 person square. Identical place settings throughout, and generous central space.
At this width, larger tables, of length 210cm/*84"* and over, are particularly suited to private domestic use.

TABLES AND CHAIRS
Separate tables: round, seating 2–8

2.03

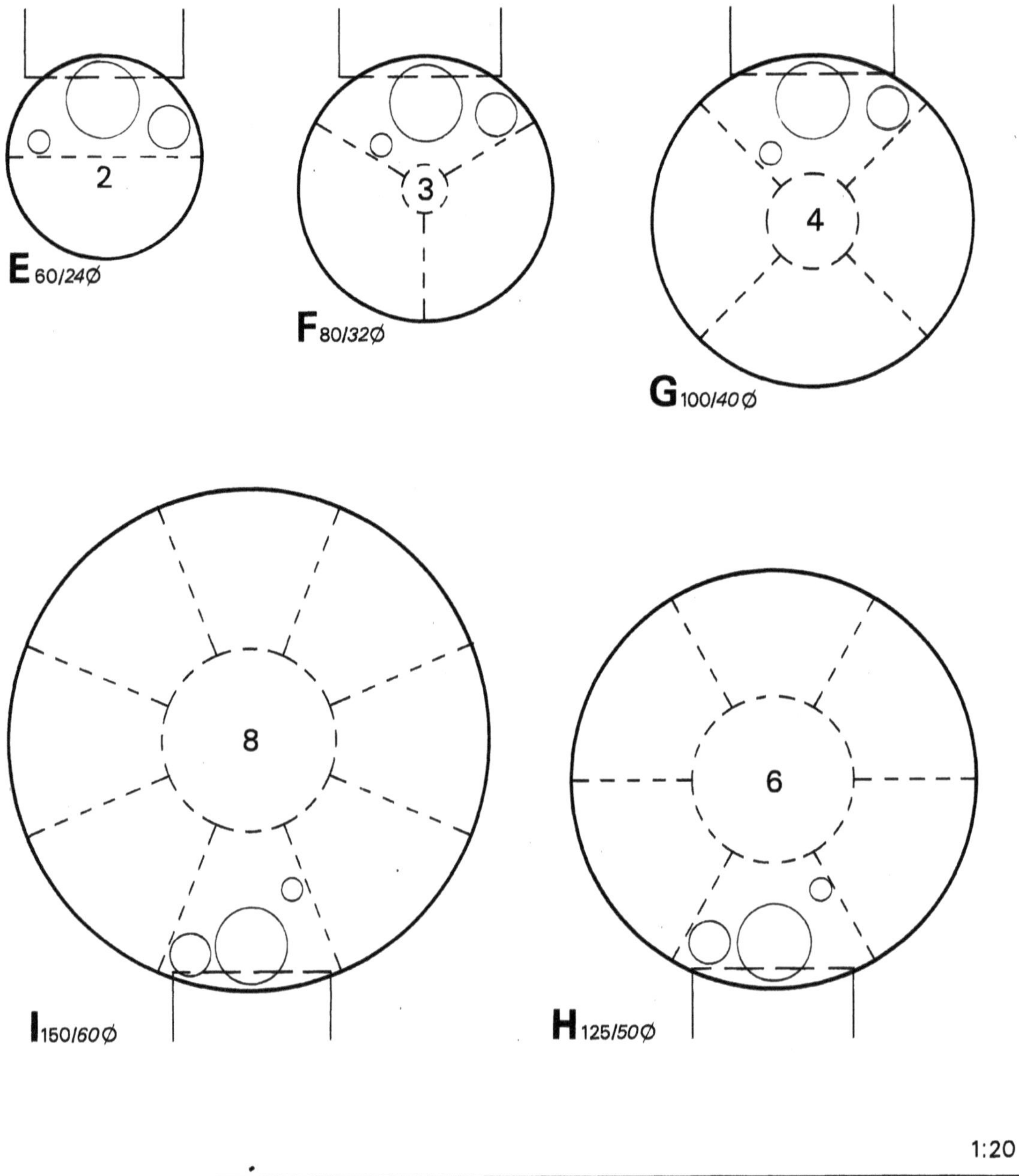

E–I Appropriate table sizes for 2, 3, 4, 6 and 8 persons.

- Straightforward range, sized in accordance with previous principles.
- Adjacent chairs, if substantially wider than 50cm/*20″* could however clash if pulled right up to the 6- and 8-seat tables.
- Semi-circular versions can clearly be derived from G, H and I. Though of marginal use, these may be appropriate to abut a viewing balcony or glazed wall.

TABLES AND CHAIRS

Separate tables: round, seating 10–12

2.04

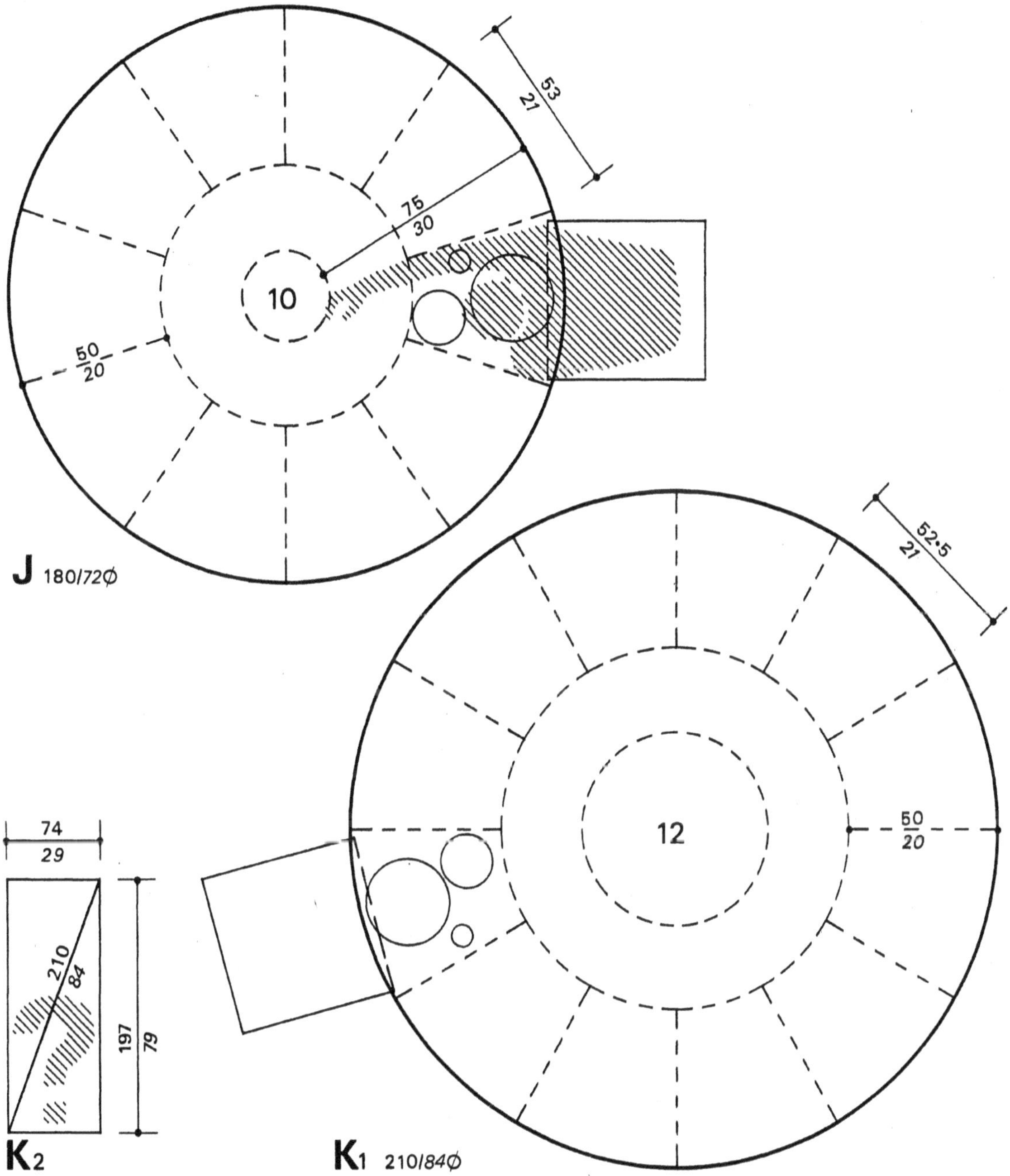

1:20, 50

J Place settings for 10 persons.
K1 Place settings for 12 persons.
K2 Size of one-piece 12-seater table top in relation to a standard door.

- Place settings, inevitably reduced in edge width, compensate by being deeper. (For relative ratios of chord length and radius, as governed by the subtending angle (see App. 3).
- Note, K1 in particular, the central dead space that lies beyond the diner's reach, even when bending fully forward. It suggests an elaborate centerpiece or, indeed, a void.
- Sheer size can pose constraints on manufacture and handling. K2 raises the question of whether a one-piece 12-seater table will quite pass through a standard door. Though the legs should in any case be detachable, the top itself should more sensibly be assembled from two or four sections.

2.05

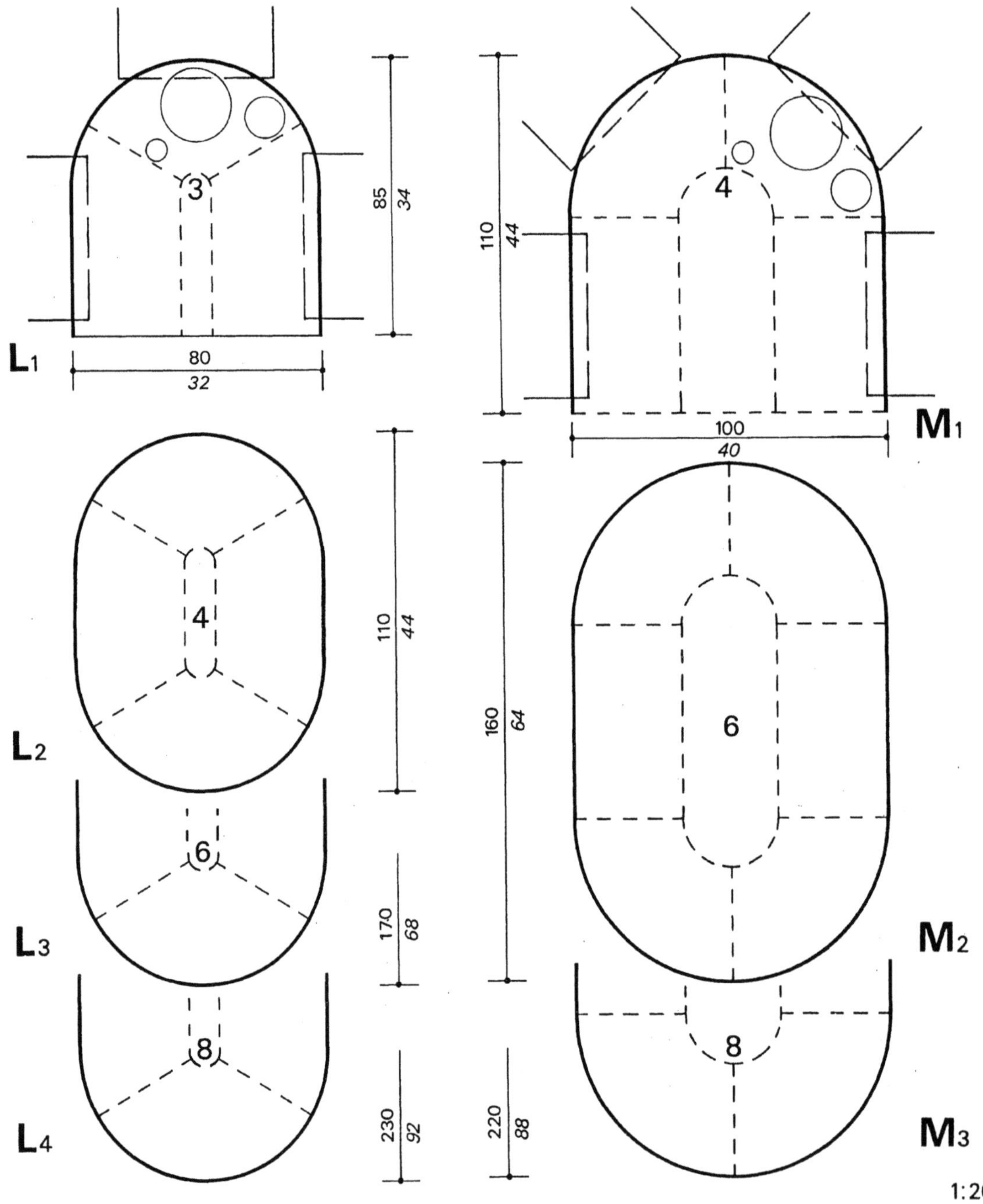

1:20

- Ovals, most commonly found in private dining rooms, are a most satisfying compromise between rounds and rectangles.
- Of the two ranges illustrated here, the L series takes a single chair at each end and the M series – which confers a more generous central surface – takes two.

L1–L4 Place settings for 3, 4, 6 and 8 persons.
M1–M3 Place settings for 4, 6 and 8 persons.

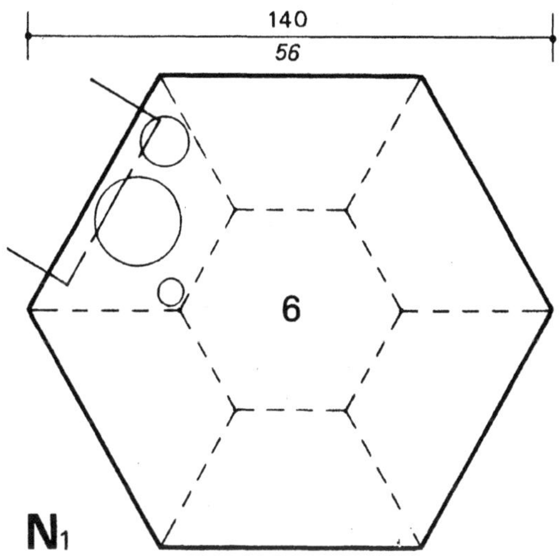

N1

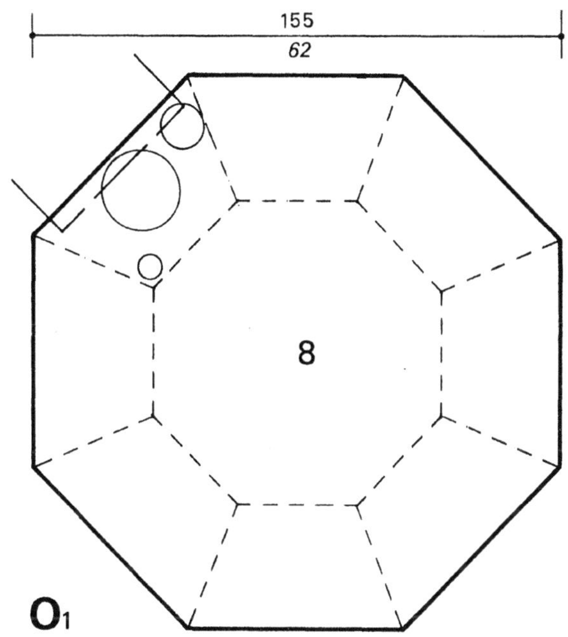

O1

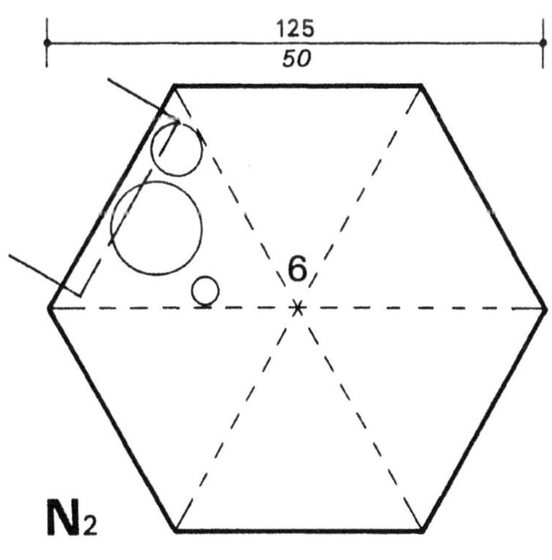

N2

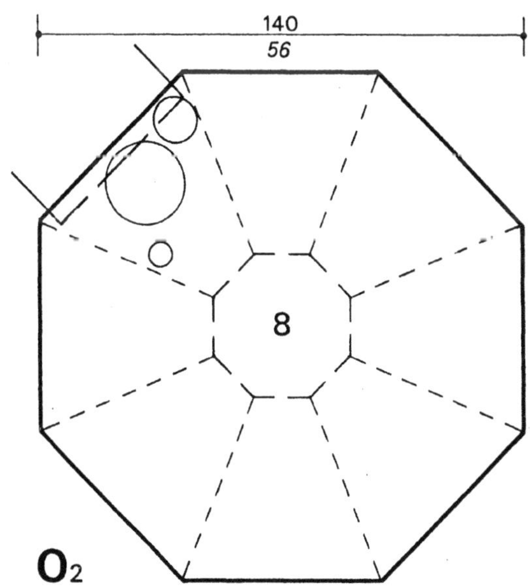

O2

1:20

N1, N2 Hexagonal tables for 6 persons.
O1, O2 Octagonal tables for 8 persons.

- Though there is not a great deal of dimensional difference between the two pairs, the examples shown in N1 and O1 give more legroom, kneeroom and center.
- The examples shown in N2 and O2 are more economic. of space and material.

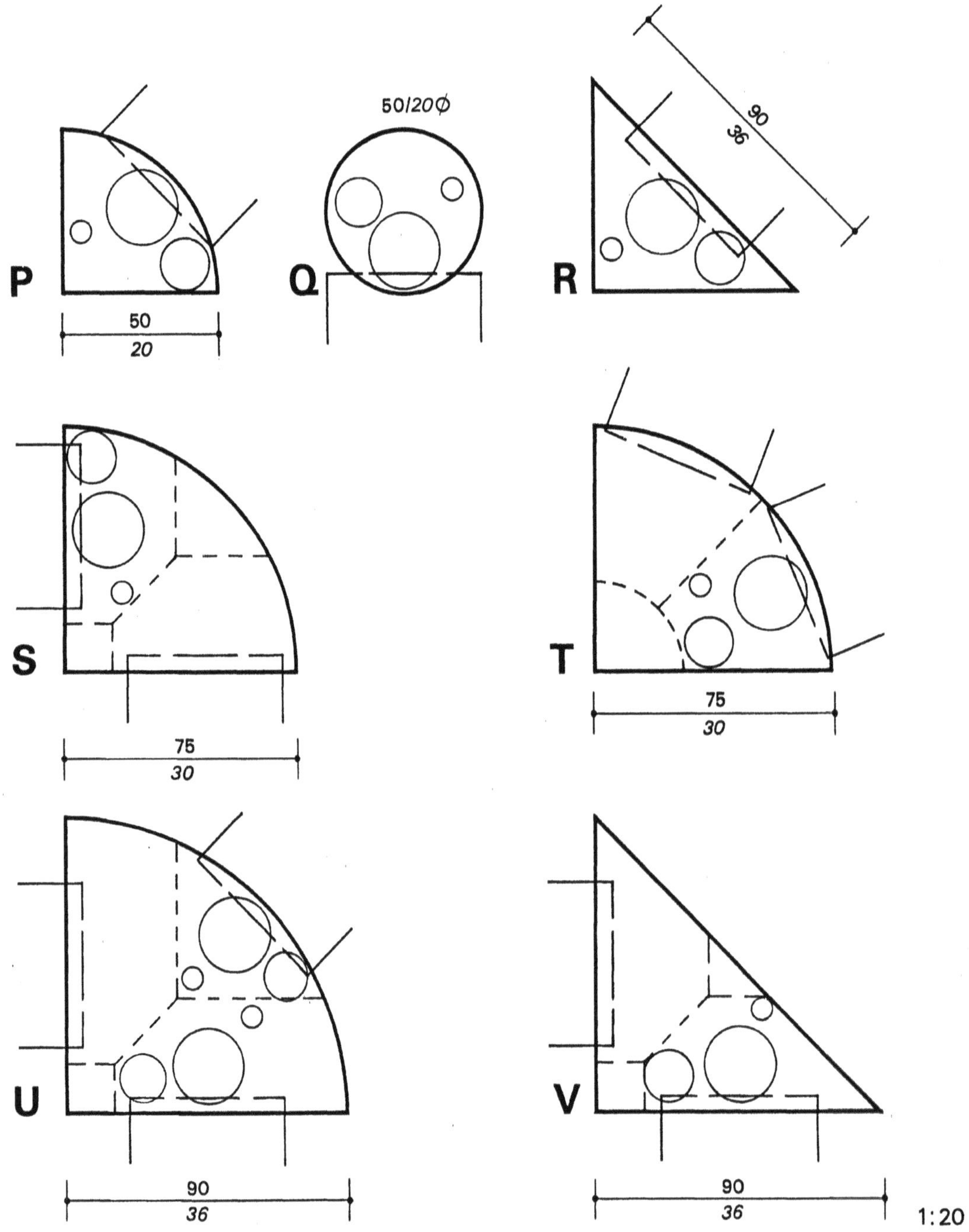

- Apart from Q – an absolutely minimal round table for one person – these small tables are all generated from full squares and circles.

P Table for 1 person: quarter of 100cm/*40″* diameter round.

Q Table for 1 person: full round.

R Table for 1 person: quarter of 90cm/*36″* square.

S Table for 2 persons: quarter of 150cm/*60″* diameter round.

T Alternative seating arrangements for 2 persons; quarter of 150cm/*60″* diameter round.

U Table for 3 persons: quarter of 180cm//*72″* diameter round.

V Table for 2 persons: half of 90cm/*36″* square. Of some value for odd corners, especially in conjunction with bench seating, and cluster formations.

TABLES AND CHAIRS
Separate tables: support structure

2.08

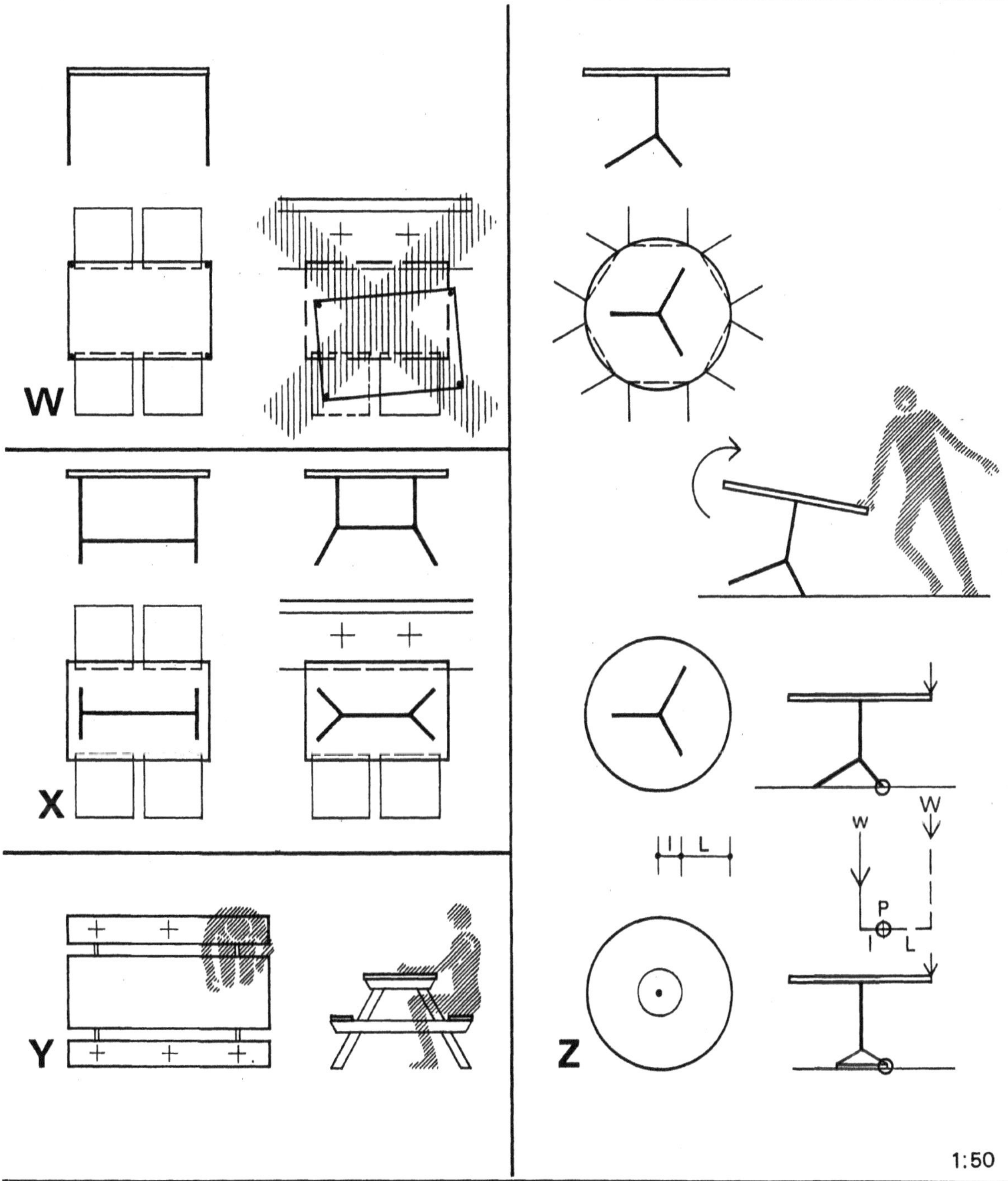

The way in which tables are supported raises three important issues – seat access, legroom and stability.

W If rectangular tables with corner legs are used with benches, the table must be moved out for anyone to get into or out of their seat: inadvisable, particularly, should the need arise during the soup course.

X Preferred alternatives: cantilevered table tops on inboard supports.

Y The cantilevered A-frame picnic table avoids the access problem by allowing the occupants to step over the bench, and even to straddle the leg frames.

Z Round pedestal tables suffer no access and legroom constraints but are potentially unstable if inadvertently leant upon. What can happen, with two types of pedestal base – splayed tripod and weighted drum – is graphically explored:

The force of the stumbler's arm (W), acting at a certain distance (L) from the pivot point (P), is countered by the weight of the table (w) acting at *its* distance (l). If the latter moment exceeds the former, the table will not tip.

Two guidelines that will reduce the risk of accident:

- The containing diameter of the tripod base should be at least *four-fifths* of the tabletop diameter.
- The smaller the weighted base, the heavier it must be.

TABLES AND CHAIRS

Combined/extended tables: rectangular – modular

2.09

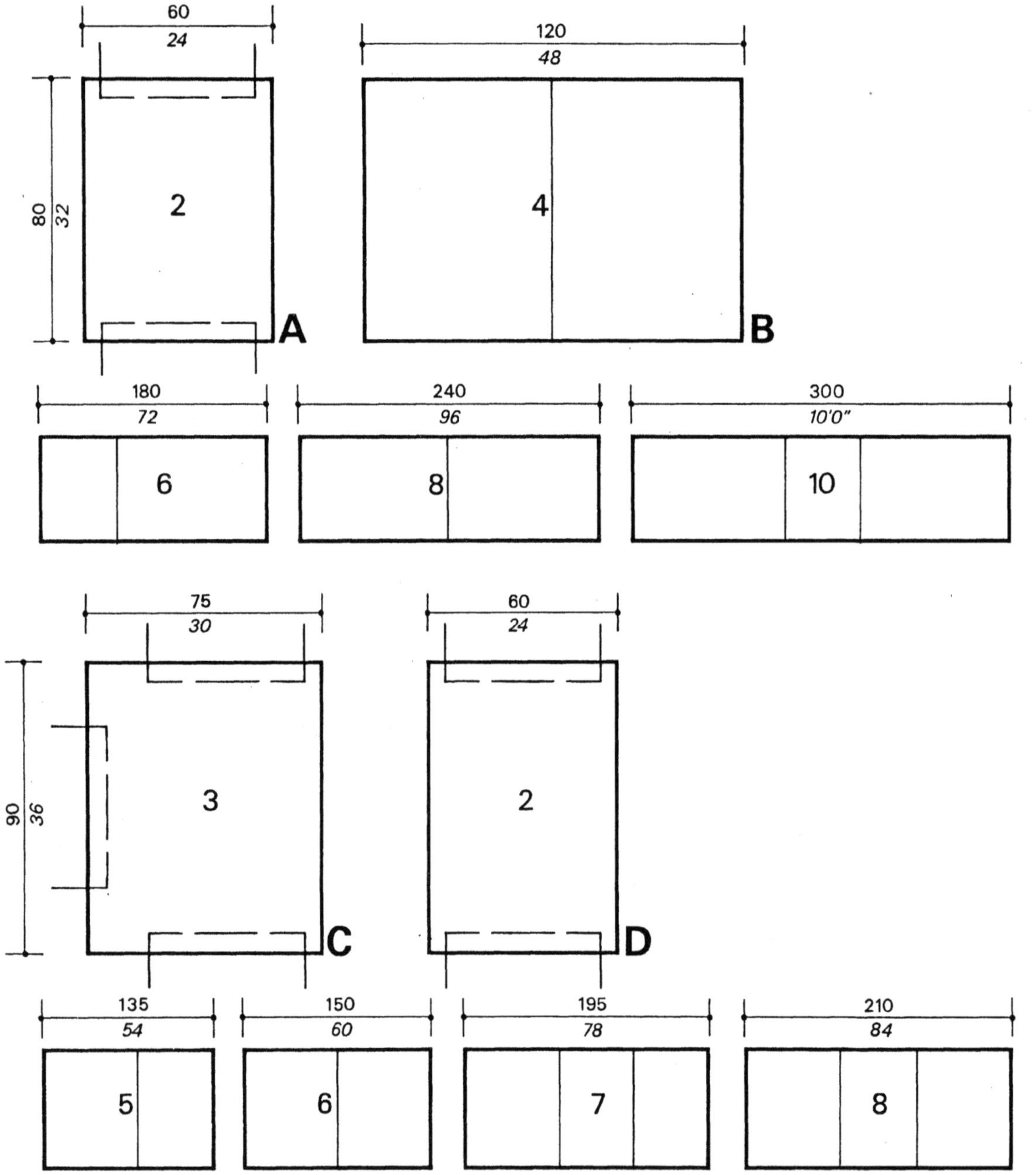

1: 20,50

A, B Arrangement of tables for 2, 3, 4, 6, 8 and 10 persons.

C, D Arrangement of tables for 2, 3, 5, 6, 7 and 8 persons.

- Two bi-modular ranges, derived from dimensions established in 2.01 and 2.02. A basic pair of easily handleable units – small free-standing tables in their own right – may be built up into a variety of longer ones, thus providing considerable flexibility of arrangement.
- Since it does not accept end-seating, the 80cm/*32"* wide series is particularly well-suited in conjunction with wall benching.

TABLES AND CHAIRS

Combined/extended tables: triangular-based – modular

2.10

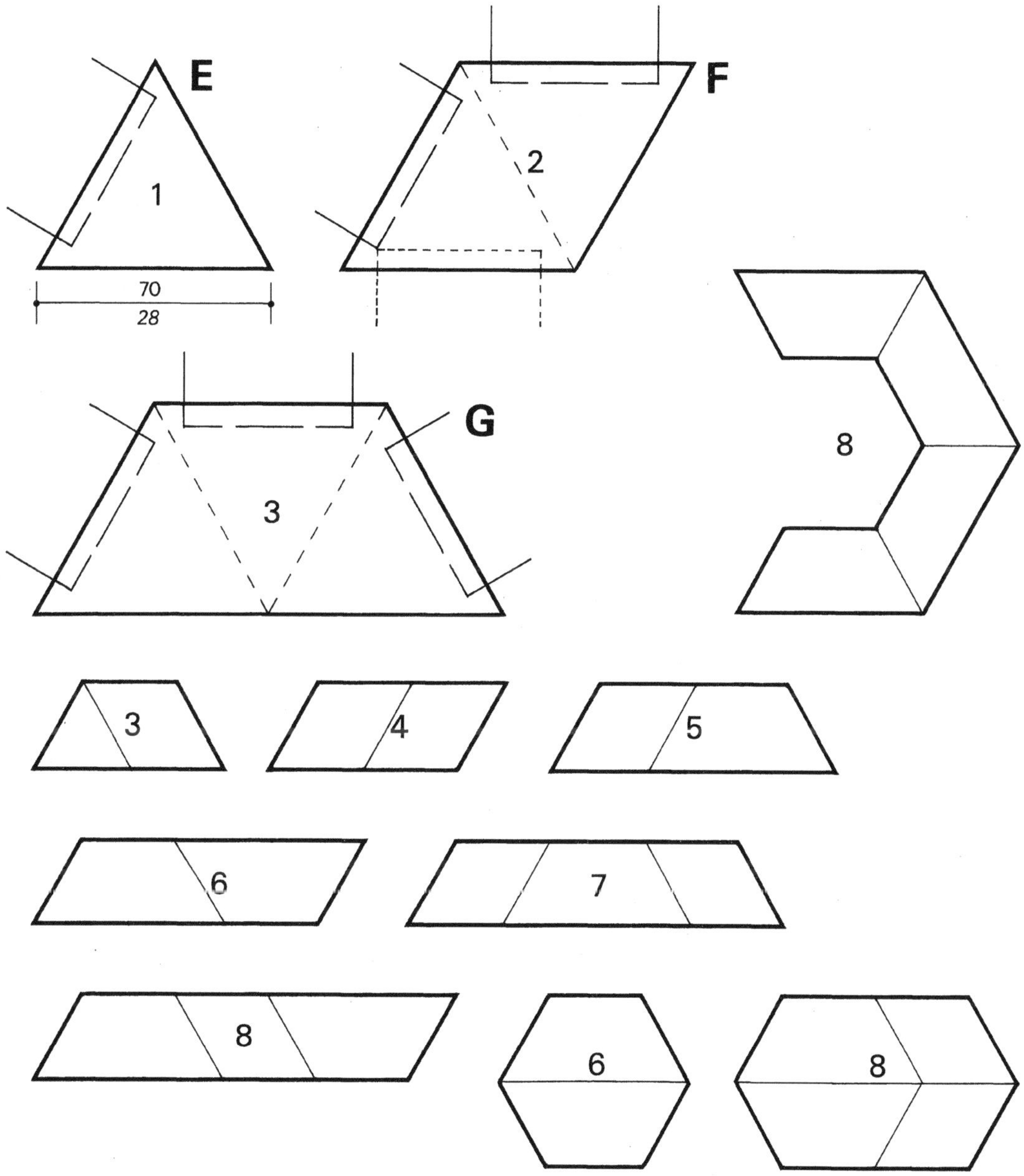

1: 20,50

E Triangular-based table for 1 person.
F Triangular-based table for 2 persons.
G Triangular-based table for 3 persons.

- Though a single equilateral table is of debatable value on its own, its diamond and half-hexagon derivatives may be endlessly combined.
- An interesting benefit, in straight run combinations, is that the triangular interlock staggers people's legs and place settings and permits a relatively narrow table width: fractionally over 60cm/*24"* in this instance.
- Note that the re-entrant "U"-shaped figure takes 8 people not 12, since any central chairs would clash.
- Such a range, with its ability to twist and turn through 60°, is most aptly deployed in any space planned with an equilateral grid.

TABLES AND CHAIRS
Combined/extended tables: square to round, seating 2–6

2.11

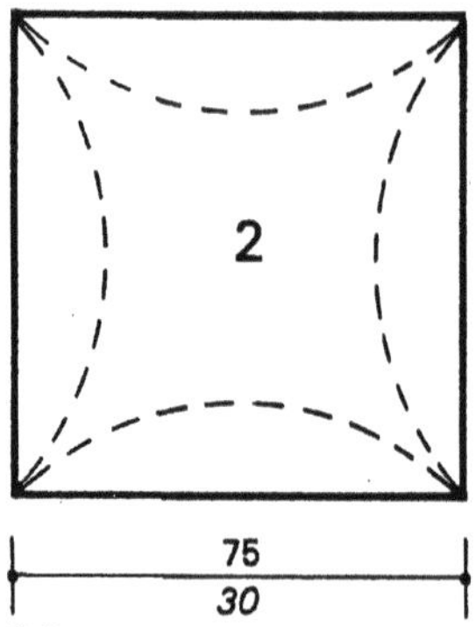

H1

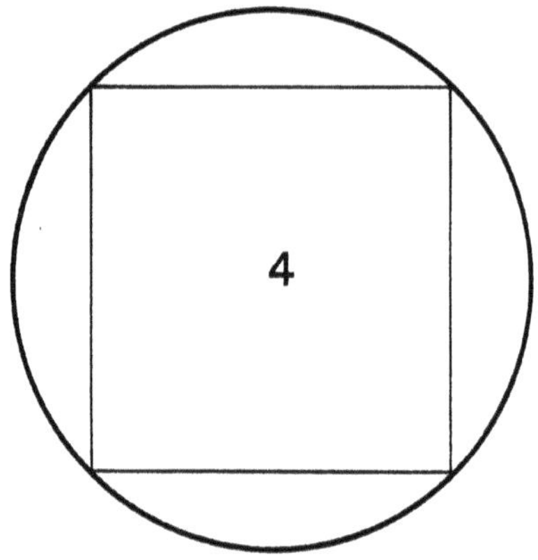

H2 105/*42*Ø

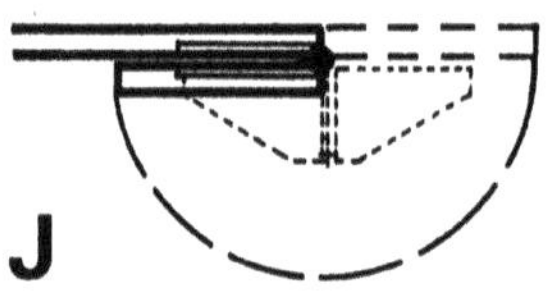

J

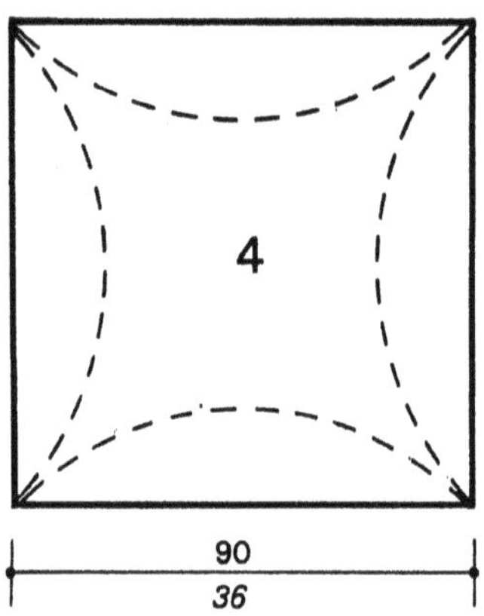

I1

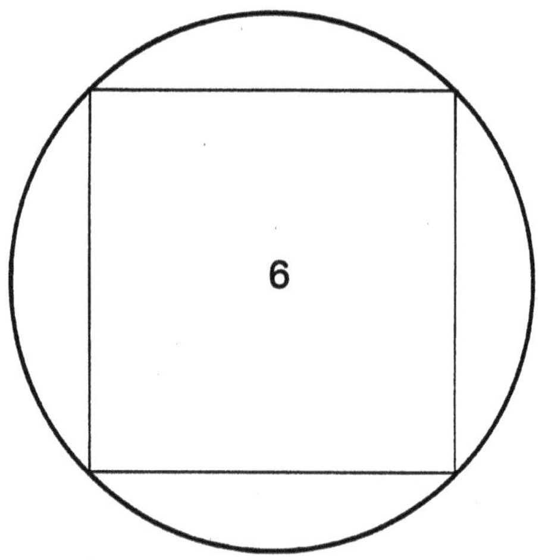

I2 127·5/*51*Ø

1:20

H1 Square table for 2.
H2 H1 converted to a round table for 4.
I1 Square table for 4
I2 I1 converted to a round table for 6.
J Hinge.

- The miracle of geometry, together with load-bearing hinges, enables a smallish square table for 2 to be converted to a generous one for 4, and a standard 4-seater to a table for 6.
- The patented hinge (J) that accomplishes this feat is remarkably ingenious. Composed of twinned pairs of steel plates, connected by powerful opposed springs, one pair flips down vertically the moment the other set has been rotated through 180°. This instantaneously forms a *rigid cantilevering bracket.*
- In the folded state, the table flaps are held tight to the underside by friction catches. Altogether, a technological advance on the traditional gate leg.

2.12

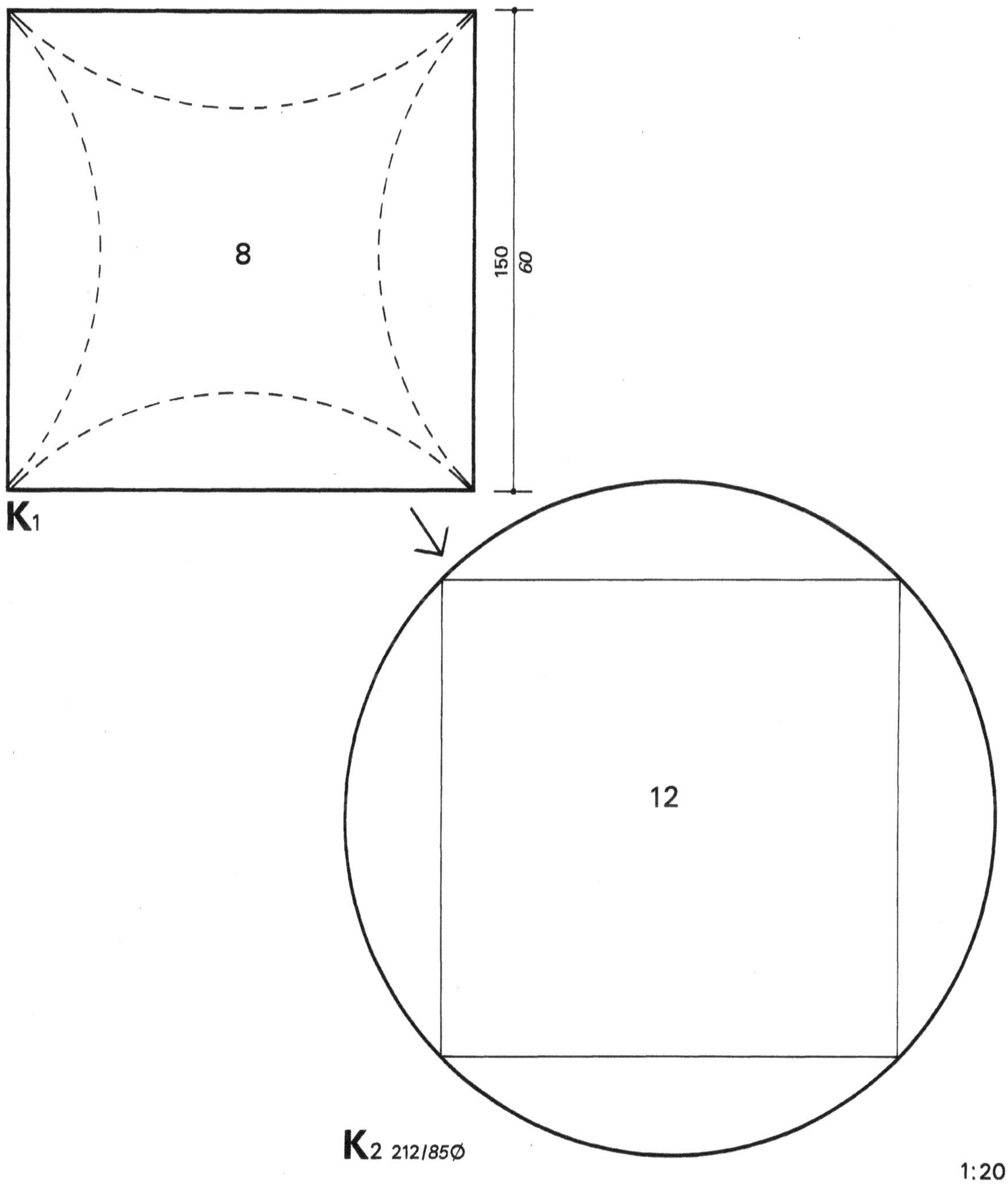

- The same mechanism, described on the preceding page, will convert a large 8-seater square to a 12-seater circle.
- With flaps folded in, the table will easily negotiate a normal door opening.

K1 Square table for 8.
K2 K1 converted to a round table for 12.

Combined/extended tables: square to square or octagon, seating 4–8

2.13

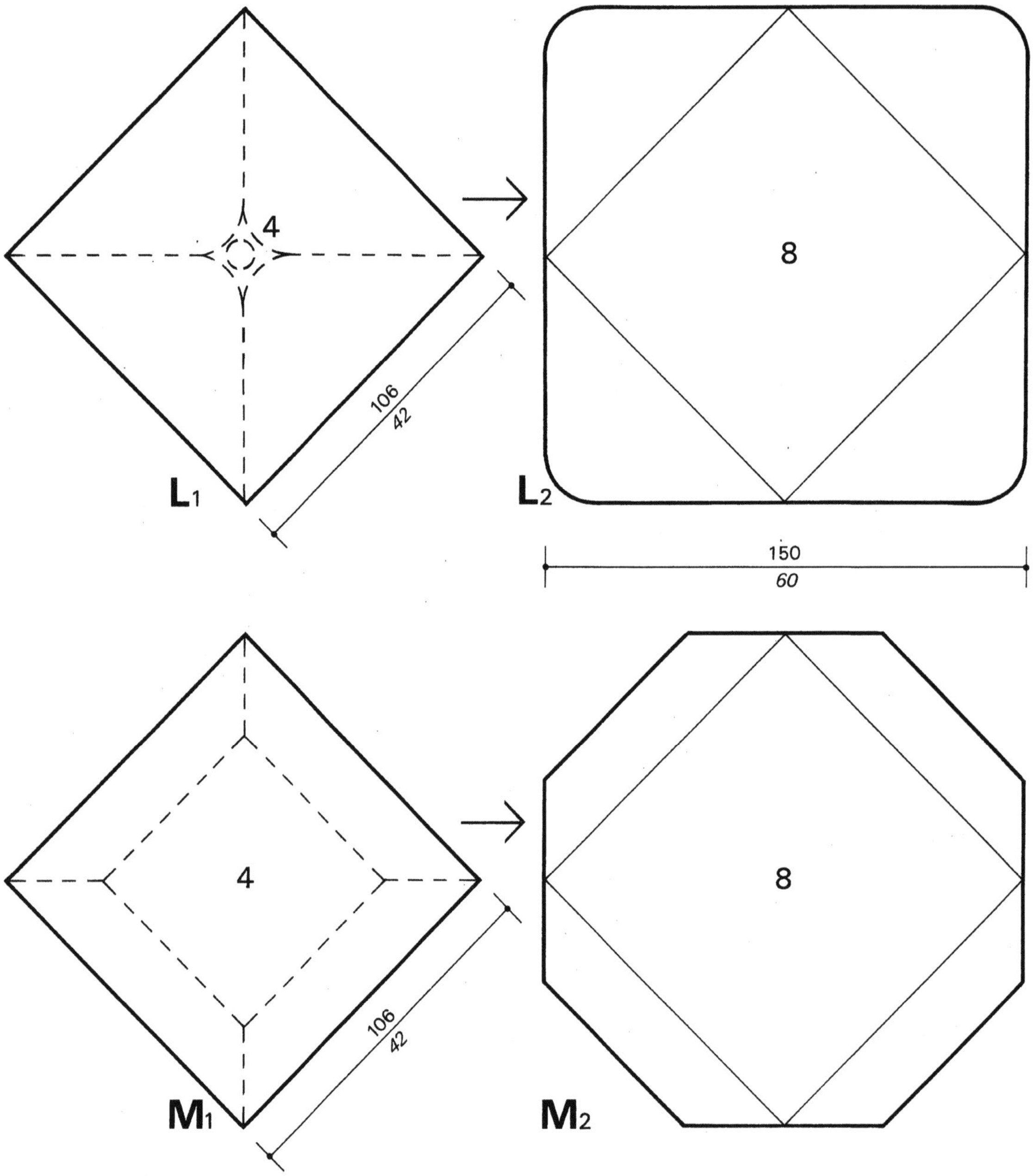

1:20

L1 Square table for 4.
L2 L1 converted to a square table for 8.
M1 Square table for 4.
M2 M1 converted to an octagonal table for 8.
Both conversions double the number of seats and are attractive further variants on the folding-flap principle.

- Whilst the dimensions of the 8-person square table are ideal, it is somewhat over-sized for 4 when retracted. Alternative side measurements, lowering the place width to 55cm/*22"*, would be approx. 99cm/*40"* expanding to 140cm/*56"*. Rounded corners – no bad thing in themselves – are a consequence of leaving space for the central pedestal.
- The octagonal version puts less demand on the hinges.
- Though not illustrated and in another context, it is worth recalling the classic card table. After rotating the top through 90°, its halves or quarters are folded out, displaying felt-like covered undersides, and supported on the underframe.

TABLES AND CHAIRS
Combined/extended tables: rectangle to hexagon or oval, seating 4–6

2.14

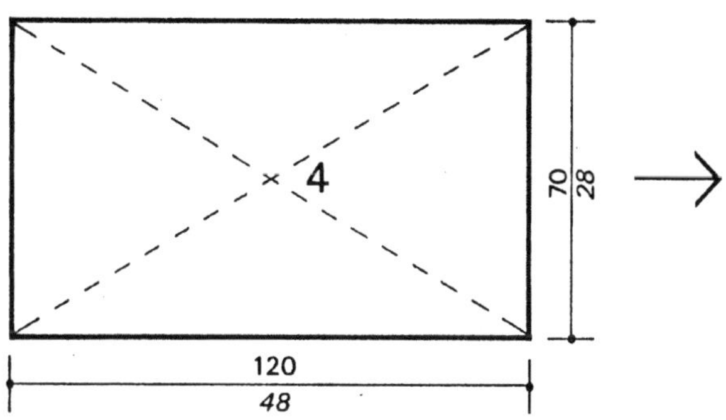

N1

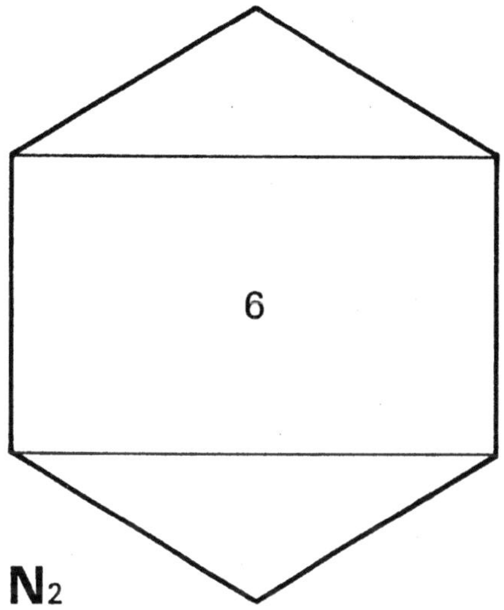

N2

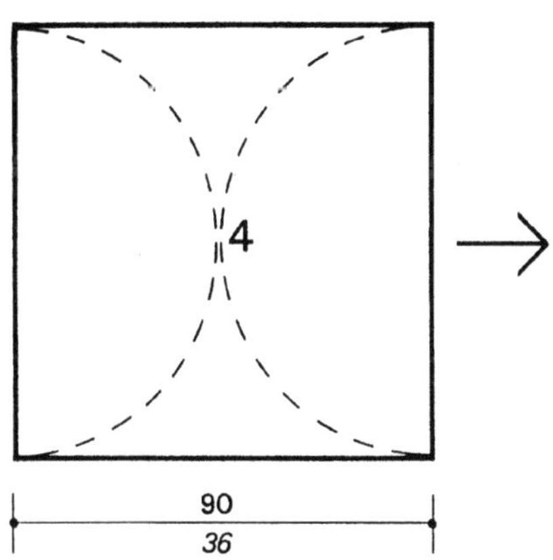

O1

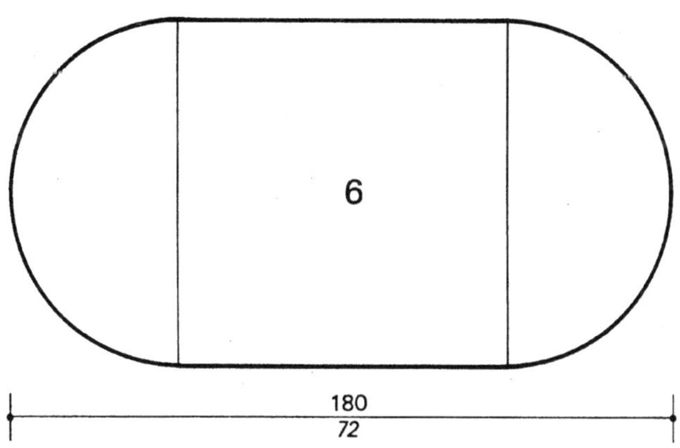

O2

1:20

- The illustrations show two elegant and commonly found extending tables.

N1 Rectangular table for 4.
N2 N1 converted to a hexagonal table for 6.
O1 Rectangular table for 4.
O2 O1 converted to an oval table for 6.

TABLES AND CHAIRS
Combined/extended tables: rectangular and round-end – linked, seating 6–14
2.15

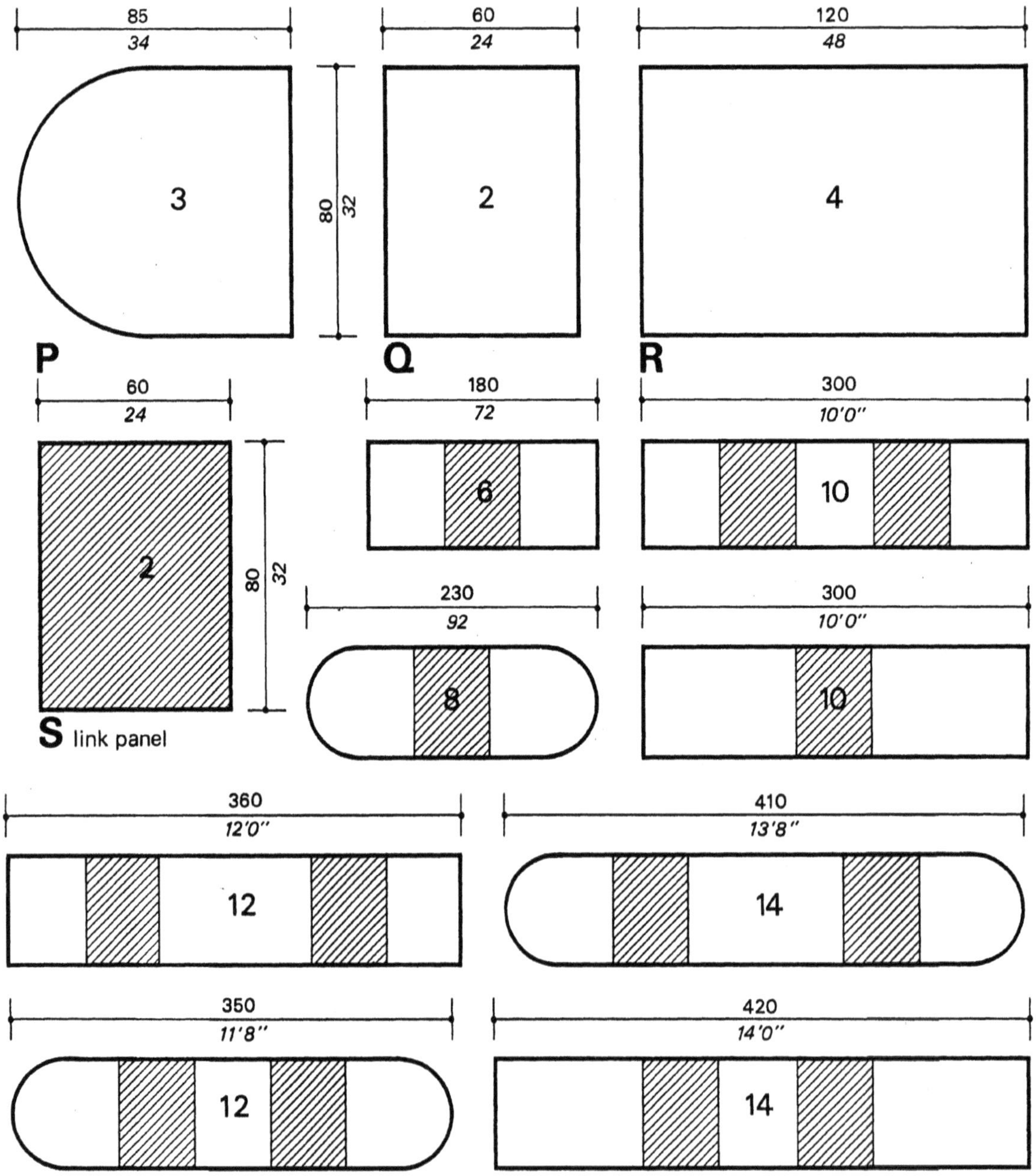

1:20,50

- A similar range to the upper ones shown in 2.09. The difference concerns the addition of a seatable round-ended unit (**P**) and incorporation of a small legless linking panel (**S**).

Combined/extended tables: triangular-based – linked, seating 10–18

2.16

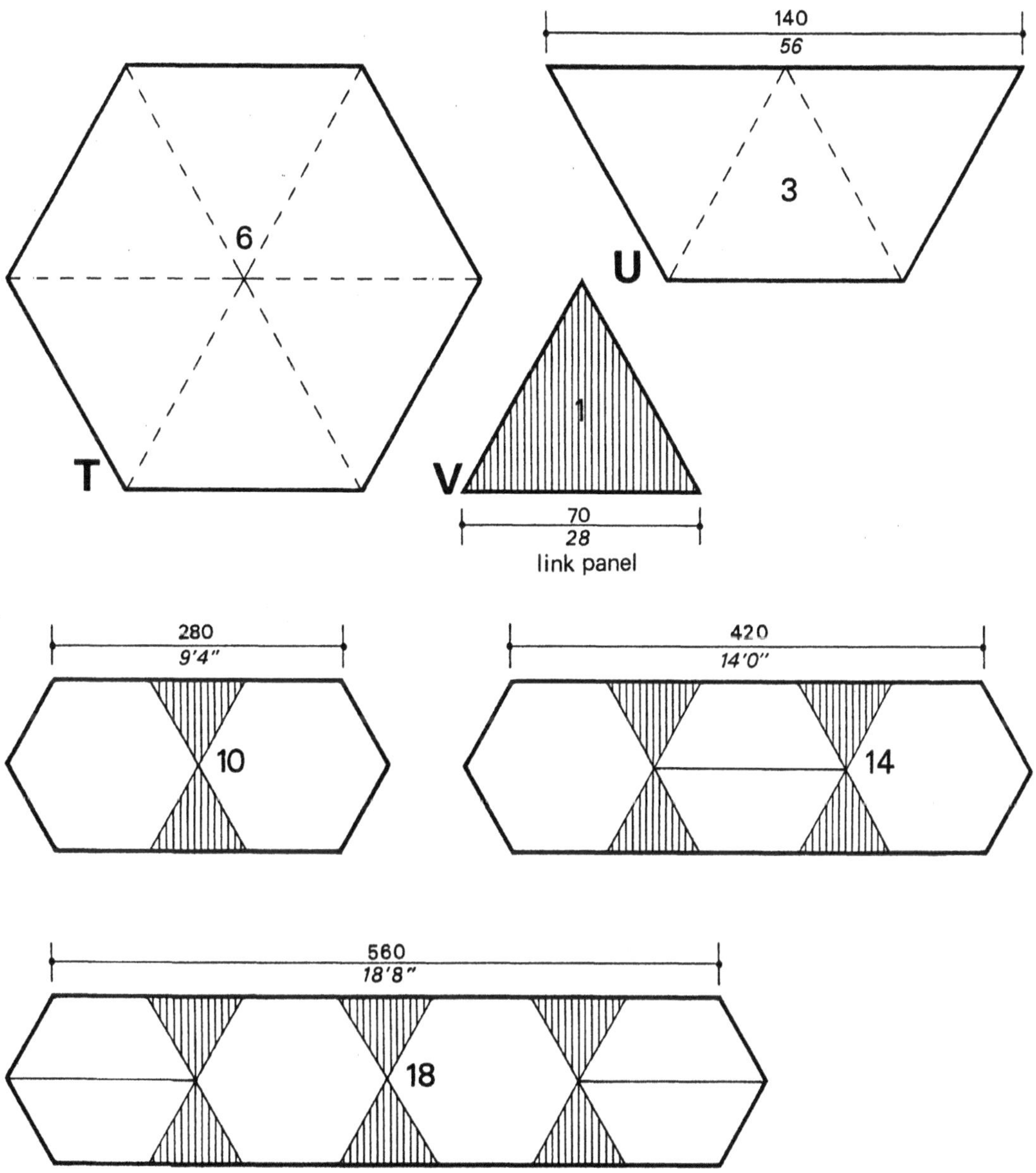

1: 20,50

T Triangular-based table for 6.
U Triangular-based table for 3.
V Triangular link for 1.

- A range of linked tables, corresponding closely to those illustrated in 2.10.
- The units are usually locked together by tongue-and-groove edge jointing.
- These and the foregoing examples, conclude this topic. Other suitable geometries will suggest themselves to the inventive designer.

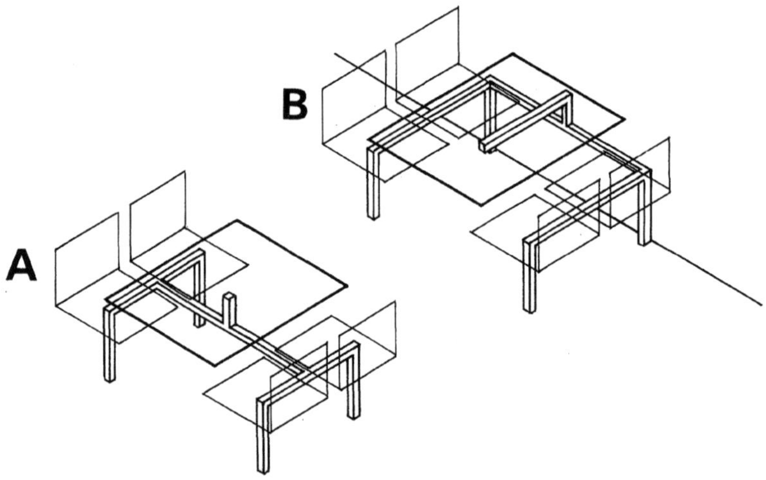

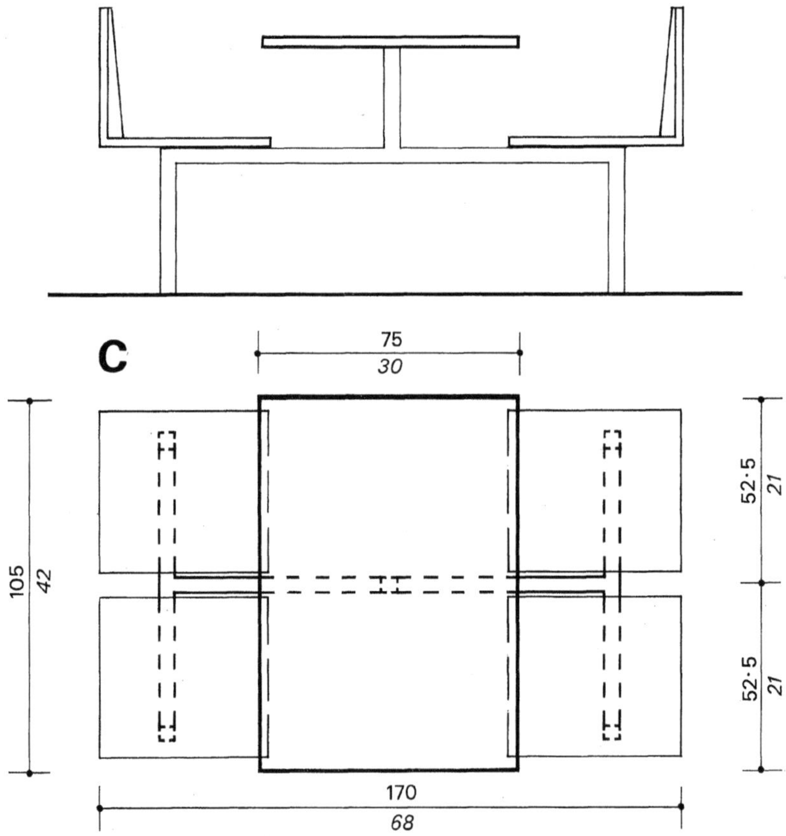

- In recent years, proprietors of cafeterias and "fast food" establishments have pioneered the use of space-saving table/seat combinations which predetermine an exact and economical layout, make minimal demands on maintenance and floor cleaning, and encourage rapid turnover.
- The essence of such furniture systems: integral seats and table top secured to a four-legged steel support frame, dimensioned to a tight module, and affording seating assemblies for 1-6 persons.
- Shown here is the archetypal 4-seater.

A "Double entry" framework: seats approached from either side.
B "Single entry" version, necessary when the unit is positioned against a wall or screen.
C Side elevation and plan. Note the typical minimal place module of 52.5cm/*21"*. This table width is just wide enough to take trays.

TABLES AND CHAIRS
Table/seat units: rectangular, seating 1–6

2.18

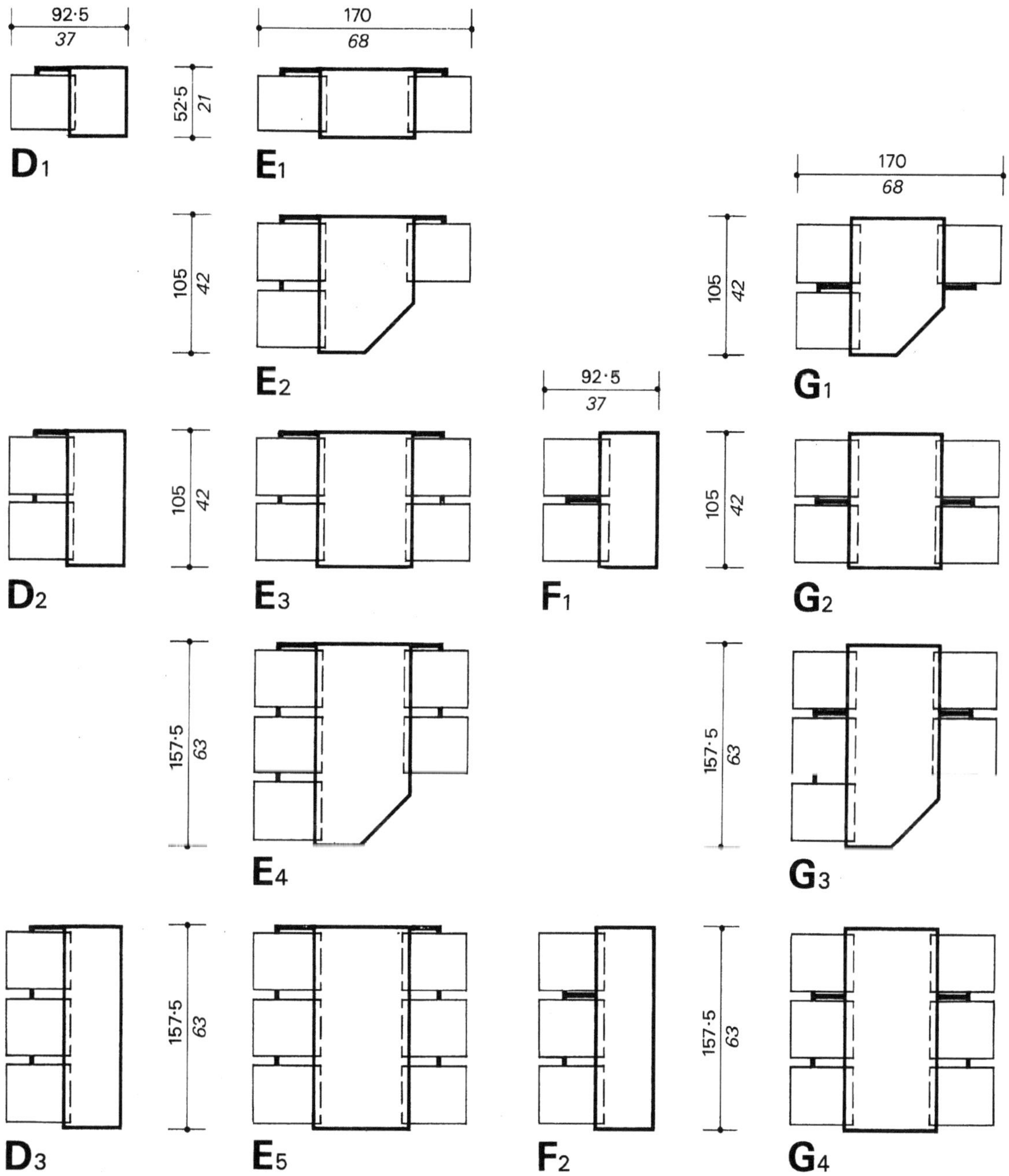

1:50

- The full potential range of 45cm/*18"* (unilateral) and 75cm/*30"* (bilateral) tables, both single-entry and double-entry, with approx. overall dimensions.
- These sizes – just adequate for cafeteria-type trays – are amongst those mass produced by a major UK supplier. However, the width of the bilateral table is often restricted to 60cm/*24"* or even less. Exceedingly cramped legroom and diminished eating surface is presumably calculated to speed the turnover yet further. Little wonder, then, that "eat and run" gives ground to "take-away".
  Note that the corner-chamfering of the 3- and 5-seaters saves aisle space and suggests angled layouts.

D1–D3 Single-entry unilateral units for 1, 2 and 3.
E1–E5 Single-entry bilateral units for 2, 3, 4, 5 and 6.
F1, F2 Double-entry unilateral units for 2 and 3.
G1–G4 Double-entry bilateral units for 3, 4, 5 and 6.

TABLES AND CHAIRS

Table/seat units: round, seating 4–6

2.19

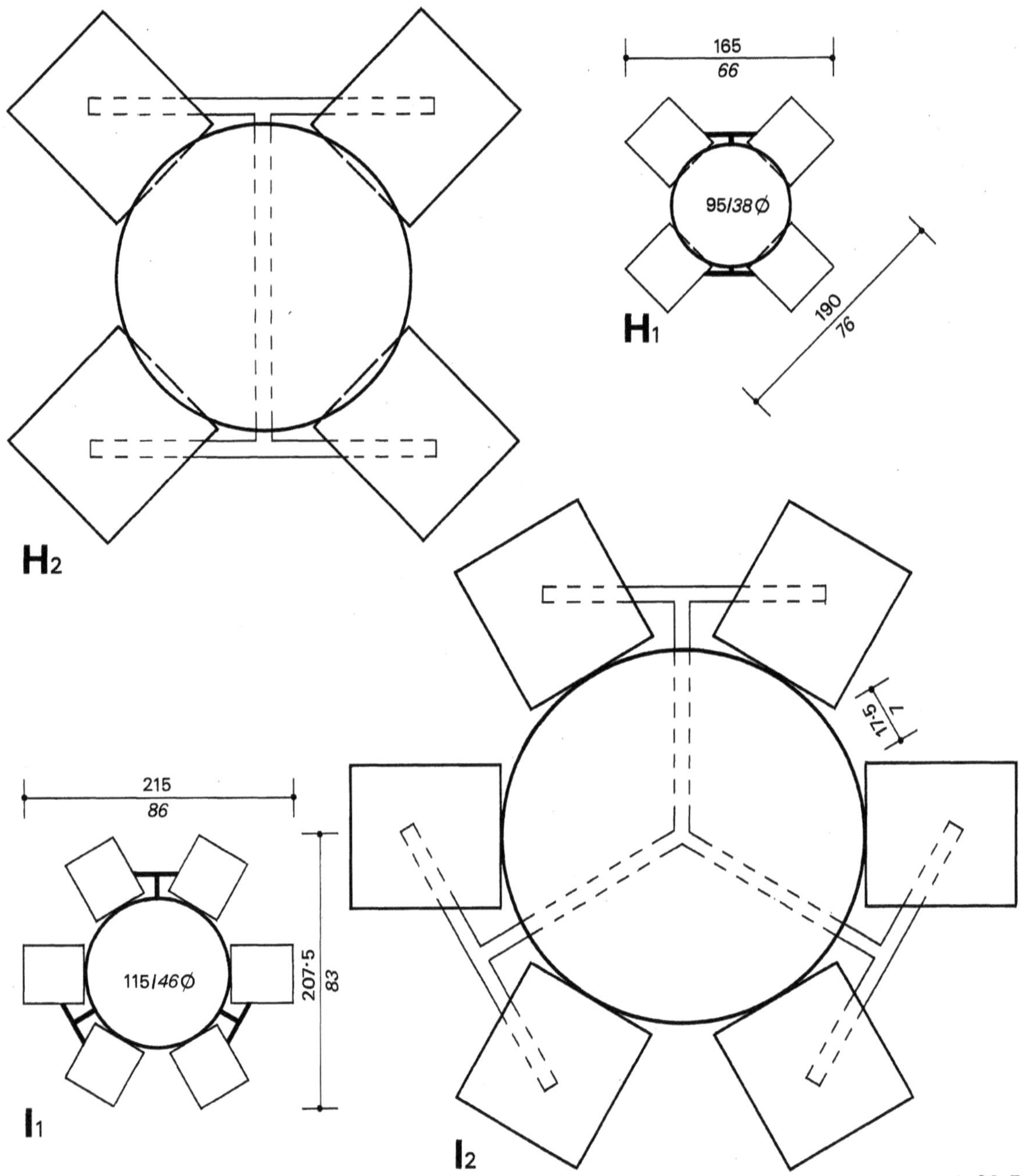

1:20,50

- Circular combinations use the same support principle illustrated in 2.18. It can be seen that customers must take their seats from one side only.
- Typical manufacturers' table sizes, 95cm/*38″* and 115cm/*46″*, are slightly smaller than the preferred diameters established on 2.03. Chairs, too, tend to be on the small side: typically 45cm/*18″* wide and 50cm/*20″* deep.

H1, H2 4-seater. A table/seat overlap, not greater than say 5cm/*2″*, is still possible.

I1, I2 6-seater. Leg access, between adjacent seats, now becomes the crucial factor. If the seats are not to be positioned further outboard, i.e. with no overlap, the front seat corners must be rounded.

TABLES AND CHAIRS

Table/seat units: rectangular, with benches, seating 2–8

2.20

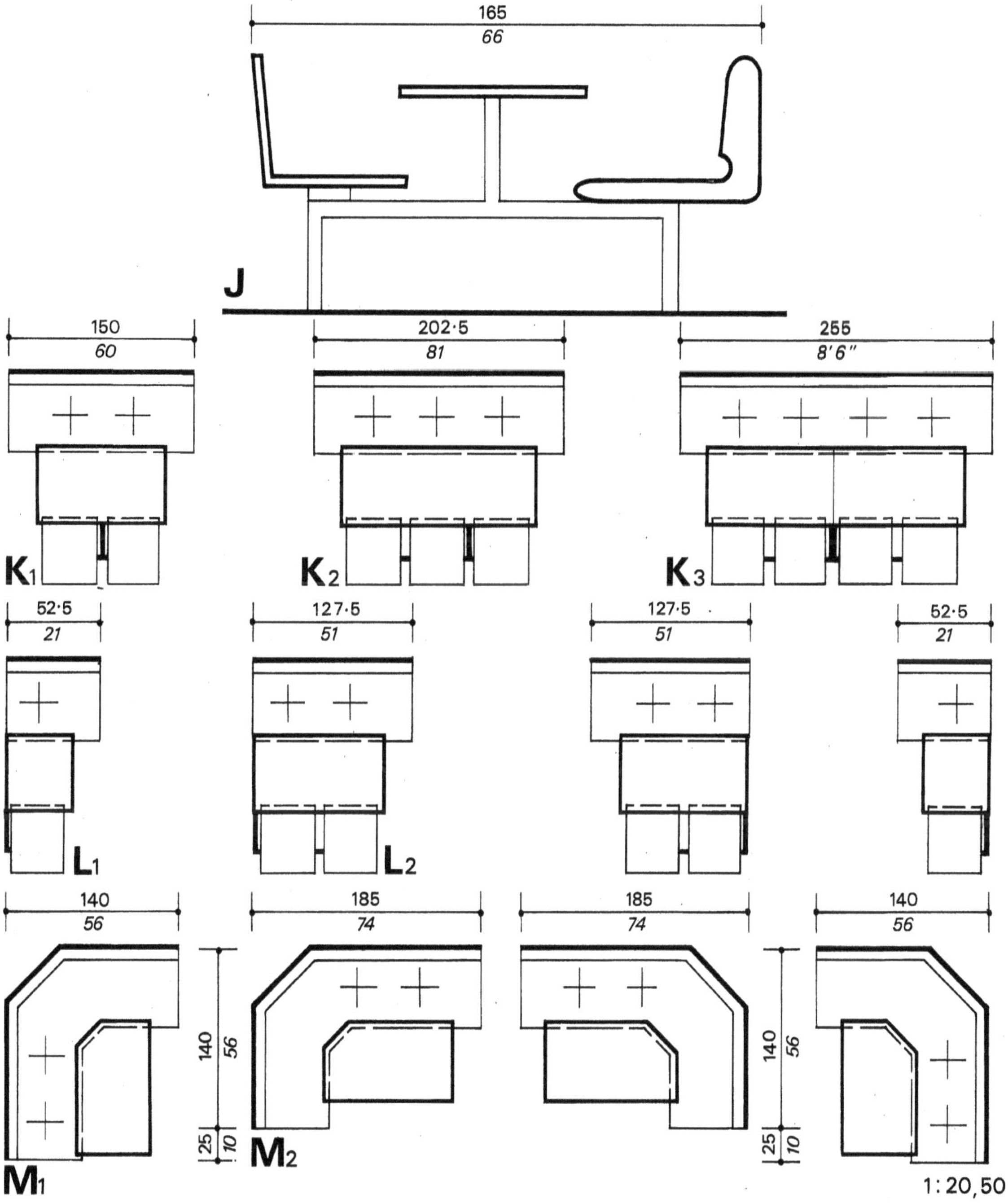

1:20,50

- The system adapted for combination with benching, constructed in all probability of fiberglass (GRP).
- Tables are shown here, quite arbitrarily, as 60cm/*24″* wide. Lengths, as before, in multiples of 52.5cm/*21″*.

J Basic section.

K1–K3 Intermediate units. Bench runs extend 22.5cm/*9″* beyond the table ends, to give 45cm/*18″* access gaps when connected.

L1, L2 End "single entry" units.

M1, M2 Corner units, seating 2 (possibly 3) at a loose table, the bench being sufficiently stable on its own. Such corner units will terminate straight runs or may permit further extensions beyond the corner.

TABLES AND CHAIRS: summary

1 The size of rectangular, round, oval, hexagonal, octagonal or trapezoidal tables – for a given number of persons – is primarily determined by the diner's need for sufficient eating surface.

2 Secondary considerations include the communal table-center space, legroom and kneeroom, the avoidance of chairs clashing, and the constraints imposed by large size on manufacture and handleability.

3 Tables with corner legs will obstruct access to bench seats, and single-pedestal tables may be unsuitable if subjected to inadvertent edge load.

4 Flexibility of arrangement and ease of handling favor the repetitive use of small modular tables, with or without insertable link-tops, that may be pushed together for large parties.

5 Strong hinge-and-support mechanisms permit a small table to become a larger one: a further example of response to varying need.

6 Table/seat units, by contrast, are purposely inflexible and compact. The emphasis, here, is on precise layouts, quick turnover, and minimal maintenance.

The next Section shifts attention to bench seating, as an alternative to individual chairs.

BOOTHS AND BENCHES

3

BOOTHS AND BENCHES
Traditional four-seater booth

3.01

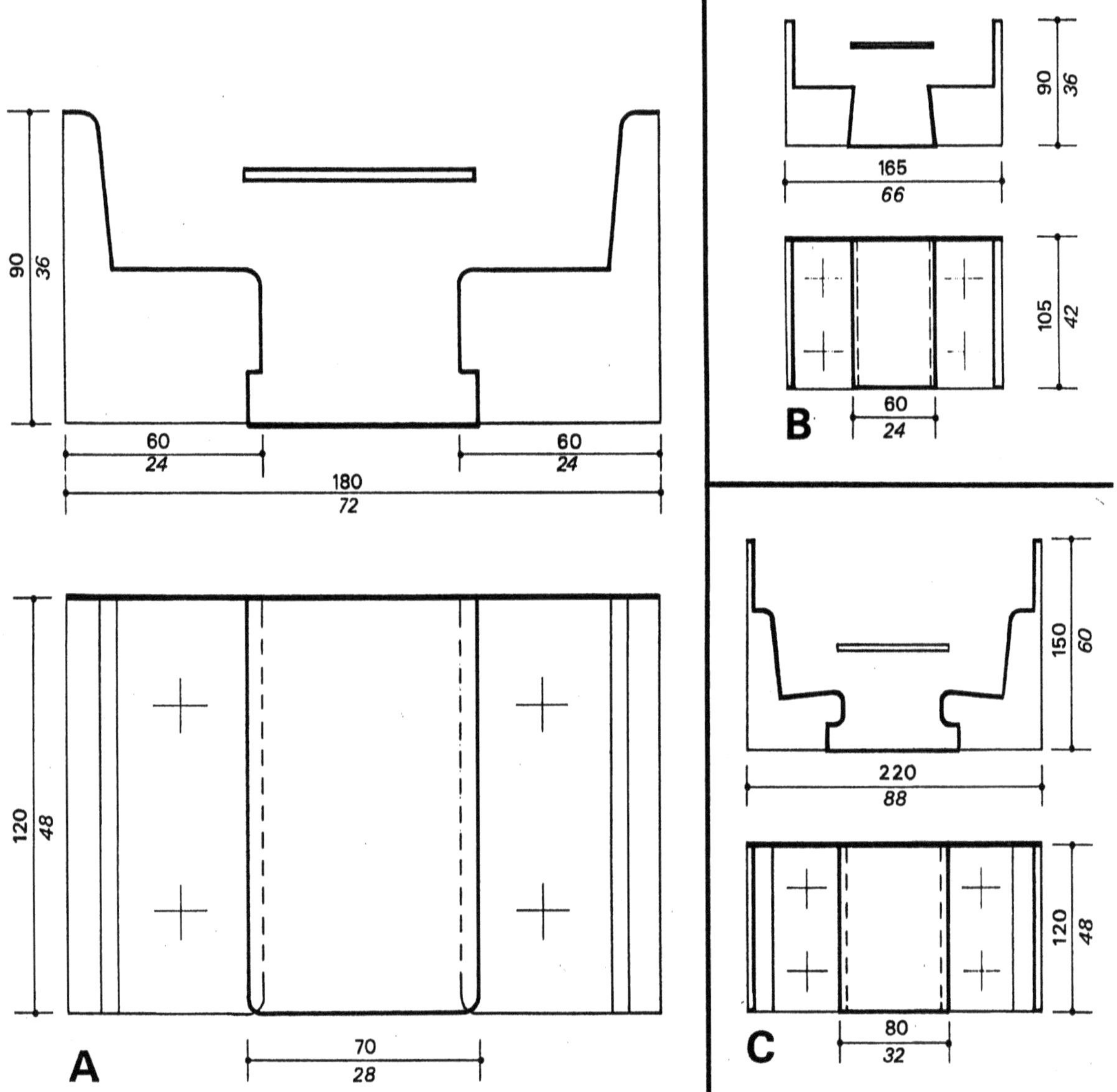

1: 20,50

- Eternal and ever popular mainstay of North American bars, drug stores, diners, highway drive-ins and downtown restaurants. Two-seaters are also common.
- Deeper booths, seating 3 or even 4 a side, are not unknown, but suffer two drawbacks. They are more difficult for innermost customers to get in and out of, and their food or drink is beyond the server's reach.
- Access to seating positions must, of necessity, be by *sit-and-slide* method: upholstery should therefore offer minimal friction, and leading edges of seats and tables should be rounded.

A Typical averaged-sized booth, with 70cm/*28"* table width.

B Minimum booth: 60cm/*24"* table width, 52.5cm/*21"* "places", harsh uncovered benches.

C Luxurious booth: wider table, upholstered seating, and dividing screens that increase the feeling of privacy.

Bench seating: straight runs and turned corners

3.02

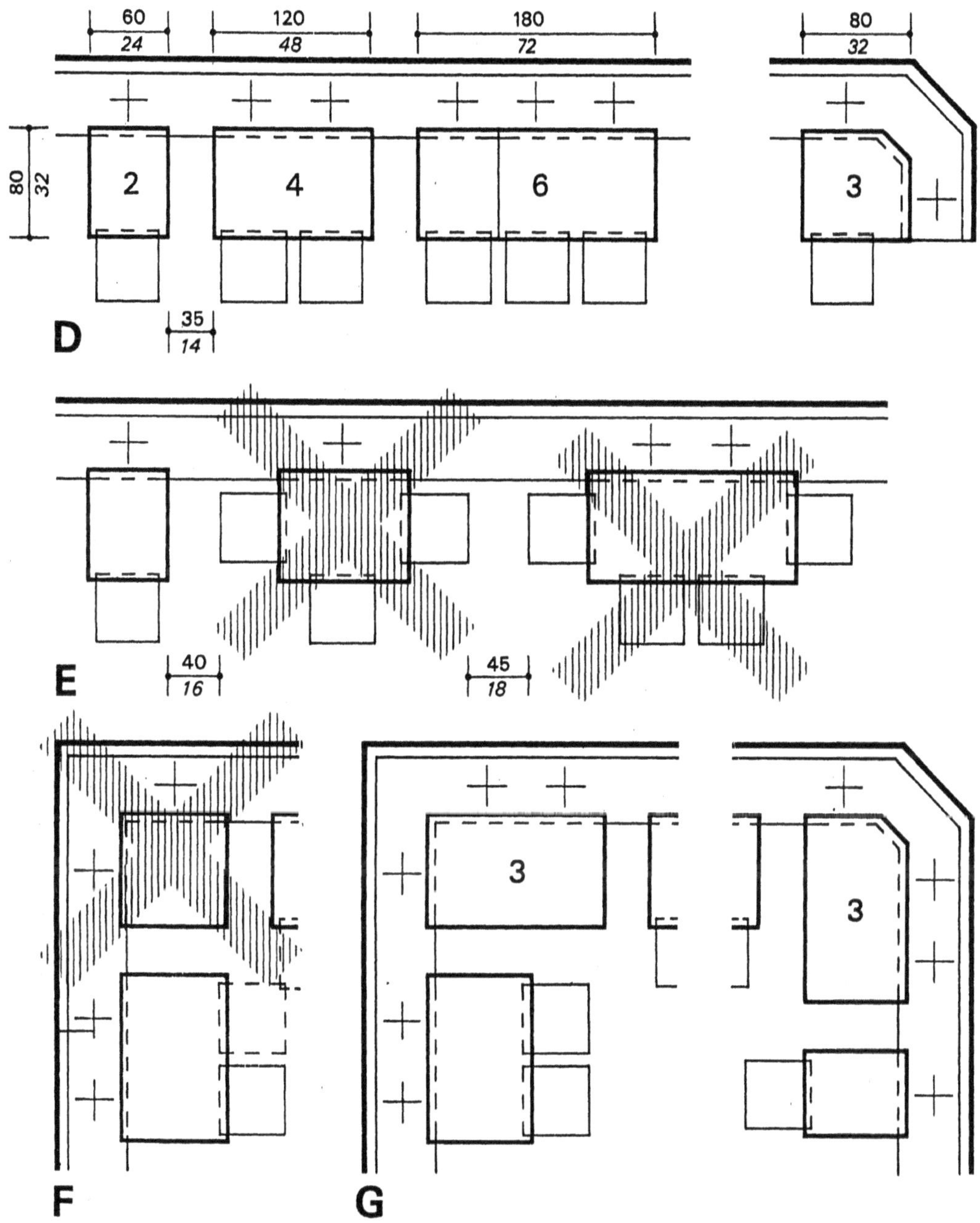

1:50

- Unlike booths, set at right angles to the wall, banquettes run alongside in continuous lengths. These upholstered benches, which afford the same comfort but less privacy, have several advantages over loose chairs:
- Continuous potential seating may be paired up with a variety of tables, which can also be joined so as to cater for larger dining groups. An essential, flexible response to changing demand.
- Accessed from the front, with tables temporarily pulled out as necessary, they save space.
- Easily turned and curved, they are suitable for a wide range of layout configurations.
- They undoubtedly enhance the overall ambience.

D Hypothetical run: tables (ideally modular) for 2 and 4, and combinable for 6 plus. Terminated, for example, by a 3-seater.

E NOT recommended. End-seating tables require large access gaps, and seat fewer in a given length.

F INCORRECT corner-turning. Such a small table will cause a clash of neighboring chairs and be inaccessible.

G Correct corner solution: longer tables for 3.

3.03

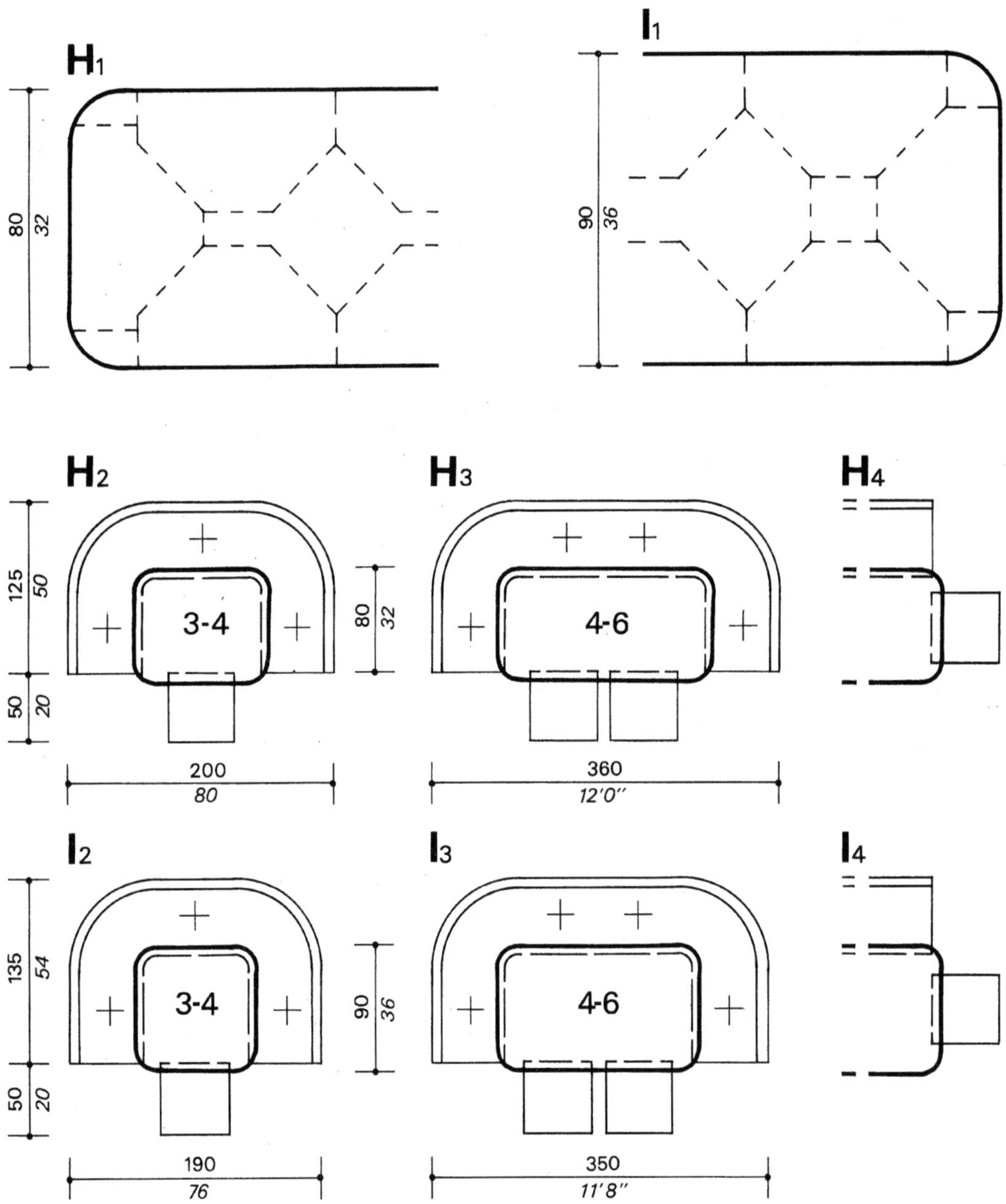

1: 20, 50

- Forming the bench seating into bays, which may well continue as a series, heightens the sense of privacy at the expense of flexibility. Undoubtedly, a most pleasant way to dine but spatially extravagant in terms of actual numbers seated and the benches themselves.
- Tables shown here are corner-rounded versions of the 80cm/*32″* and 90cm/*36″* wide rectangular range that were introduced in 2.02.
- The same geometry applies to turned-corners generally, to longer bays if required, and to any open ends.

H1	Basic rectangular table, 80cm/*32″* wide.
H2, H3	Bays, with 80cm/*32″* wide tables, for 3–4 and 4–6.
H4	Open-end treatment.
I1	Basic rectangular table, 90cm/*36″* wide.
I2, I3	Bays, with 90cm/*36″* wide tables, for 3–4 and 4–6.
I4	Open-end treatment.

BOOTHS AND BENCHES

Bench seating: curved bays and appropriate tables, seating 3–7

3.04

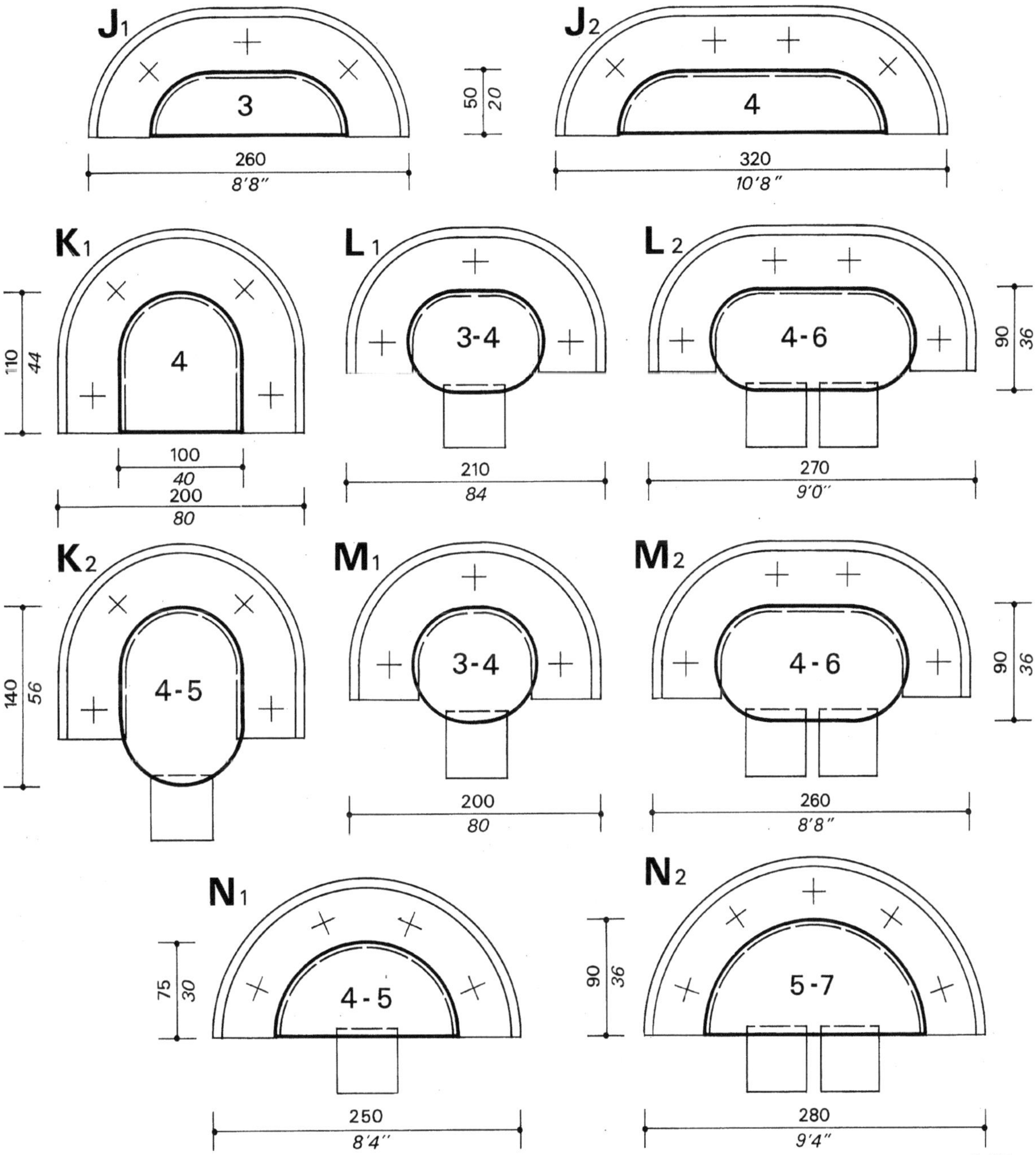

1:50

- In this instance, the half-oval and semi-circular tables are mostly as illustrated, or derived from, those shown in 2.05.
- By their nature, oval tables facilitate bench access and the optional loose chairs will not impede waiter service. Such chairs will however, as in previous examples, project into the aisle.
- Bays illustrated include some with extremely generous unilateral seating (**J1**, **J2**), suited to watching a spectacle, and a classic horseshoe shape (K1) that is effectively, an alternative booth.

J1, J2 Shallow bays, with unilateral 50cm/*20"* wide half-oval tables, for 3–4.

K1 Horseshoe bay, with 100cm/*40"* wide half-oval table for 4.

K2 Horseshoe bay, with 100cm/*40"* wide oval table, for 4–5.

L1, L2 Bays, with 90cm/*36"* wide oval tables, for 3–4 and 4–6.

M1, M2 Bays, with 90cm/*36"* wide oval tables, for 3–4 and 4–6.

N1, N2 Bays, with 150cm/*60"* and 180cm/*72"* diameter semi-circular tables, for 4–5 and 5–7.

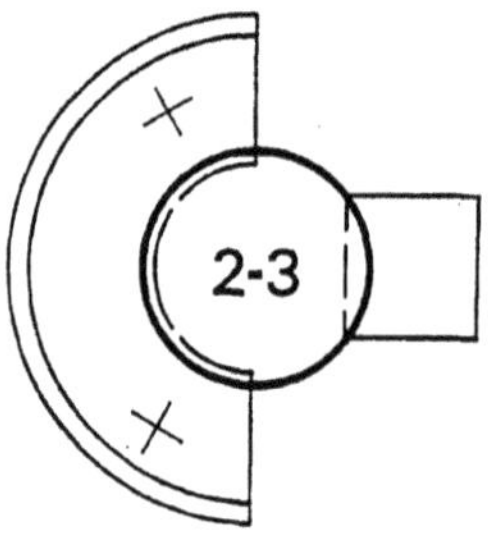

O_1 80/*32*Ø

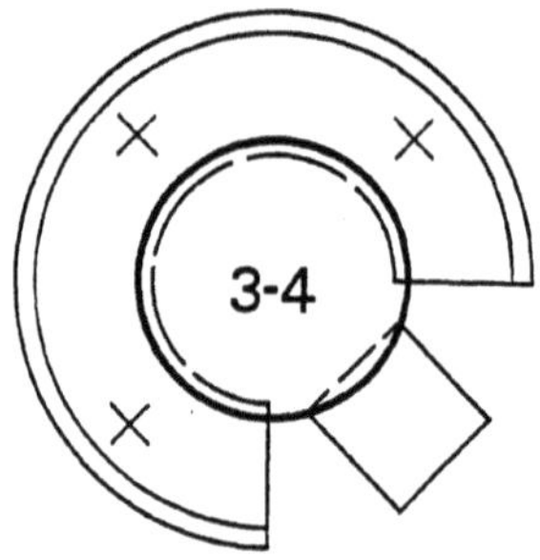

O_2 100/*40*Ø

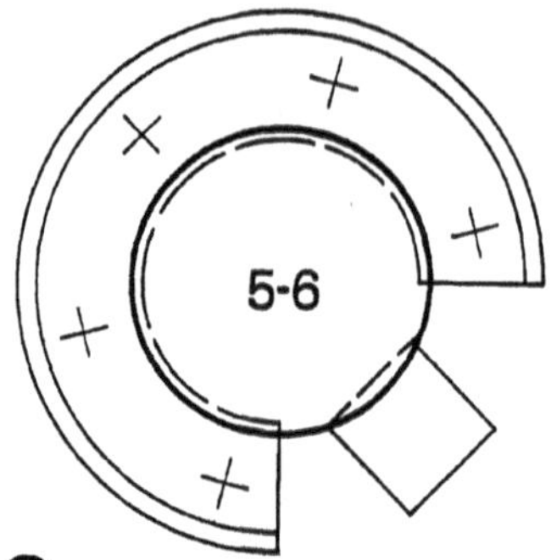

O_3 125/*50*Ø

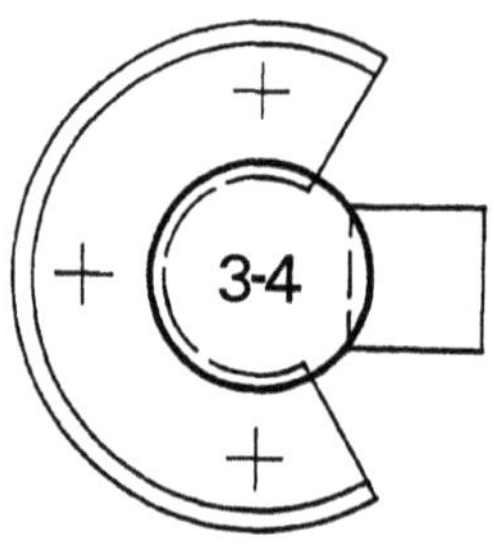

P_1

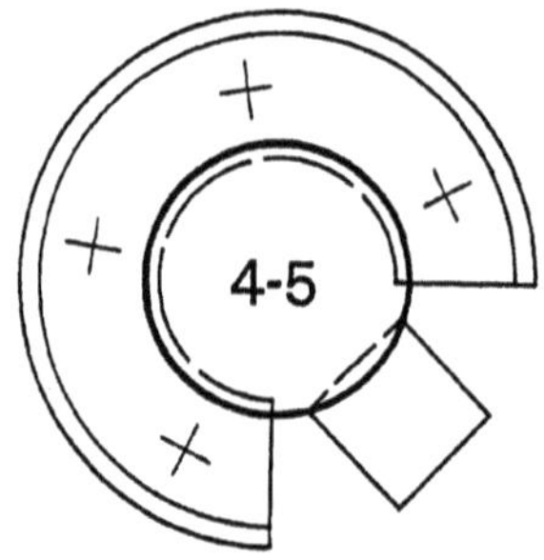

P_2

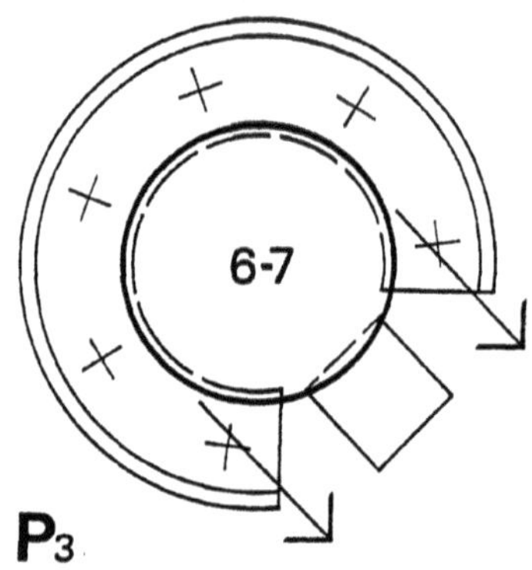

P_3

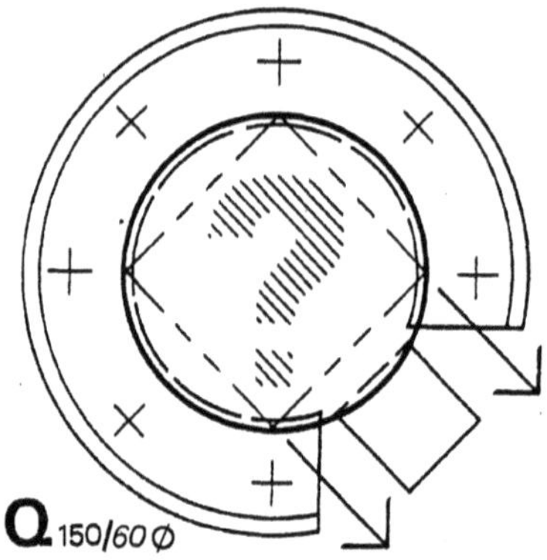

Q 150/*60*Ø

1:50

- Part-circular bays epitomize sensations of comfort and privacy. Snugness, and the number of diners seated, increase as the enclosing sweep of bench increases.
- Not without problems, however, as this escalating range demonstrates. As the ring tightens, the optional chair becomes yet more of an access and serving constraint.
- Large tables (P3 and Q) cannot be easily extricated for floor cleaning unless they have hinged flaps.
- With round-table bays there is the inherent temptation to cram in one more person. In such an event, place settings will be reduced by up to 10cm/*4″*.

O1–O3 Bays, with round tables, for 2–3, 3–4 and 5–6: generous place widths.

P1–P3 Bays, with same round tables, for 3–4, 4–5 and 6–7: minimal place widths.

Q Bay, with large round table for 7–8: extrication problem.

BOOTHS AND BENCHES

Bench seating: corner alcoves and clusters, seating 2–4 and 8–16

3.06

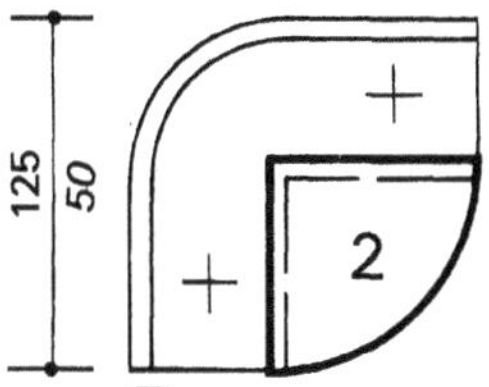

R1

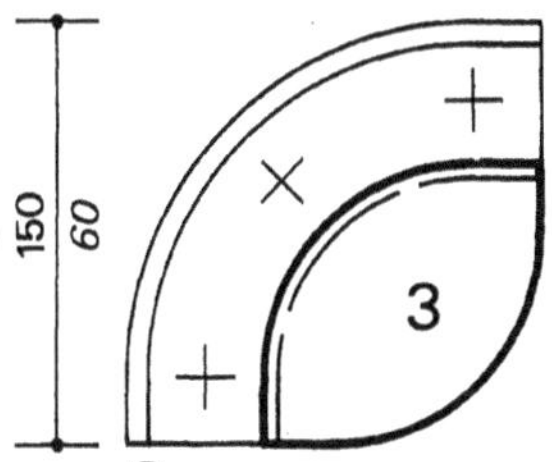

S1

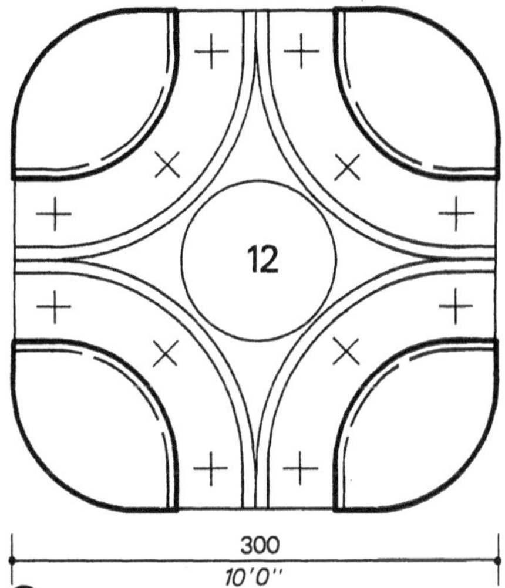

S2

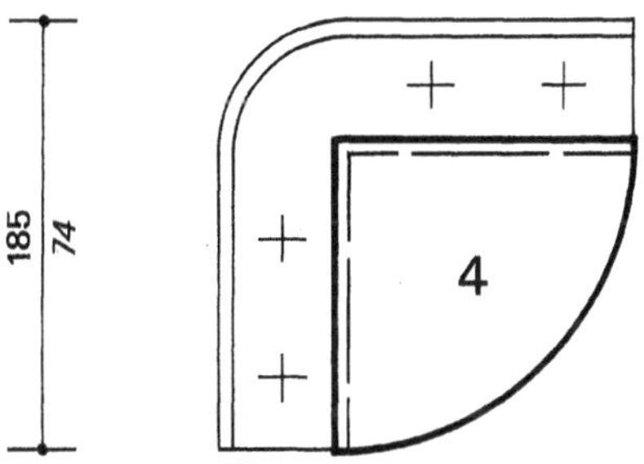

T1

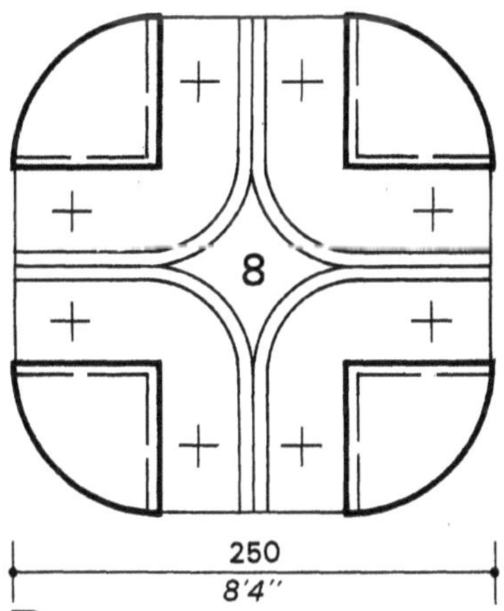

R2

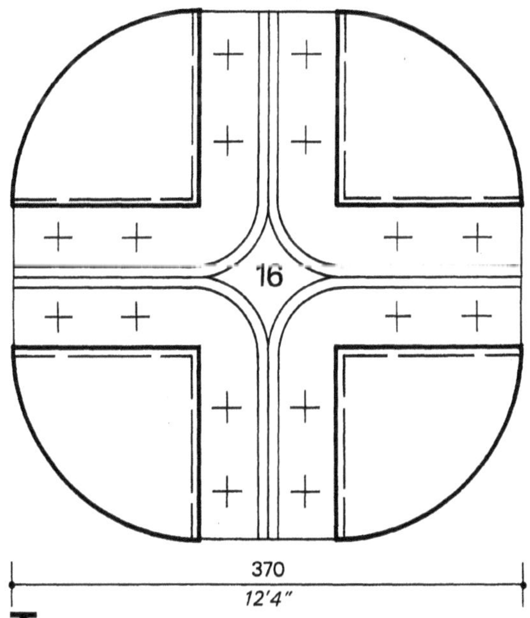

T2

1:50

- Corner alcoves (R1, S1, T1), composed of small quadrant tables (sides 75cm/*30"*, 135cm/*54"*) and of a lenticular shape that echoes the curve of the bench, may be grouped to form relatively compact free-standing clusters without added chairs.
- Where overall space permits, a single cluster can be an attractive central feature. Circulation paths will not be affected, and any structural column is neatly absorbed in the middle.
  In example S2, the cluster is capable of enclosing a more massive obstruction. Alternatively, a very large center piece is required.

R1 Corner alcove with small quadrant table, for 2.
R2 Cluster for 8 (derived from R1).
S1 Corner alcove with lenticular table, for 3.
S2 Cluster for 12 (derived from S1).
T1 Corner alcove, with large quadrant table, for 4.
T2 Cluster for 16 (derived from T1).

BOOTHS AND BENCHES

Bench seating: corner alcoves and clusters, seating 3, 4, 12 and 16

3.07

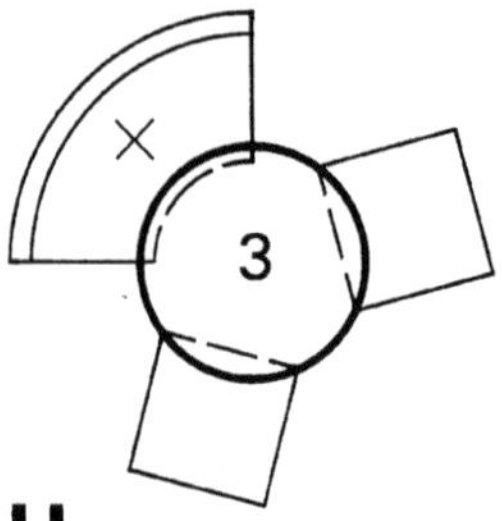

U1

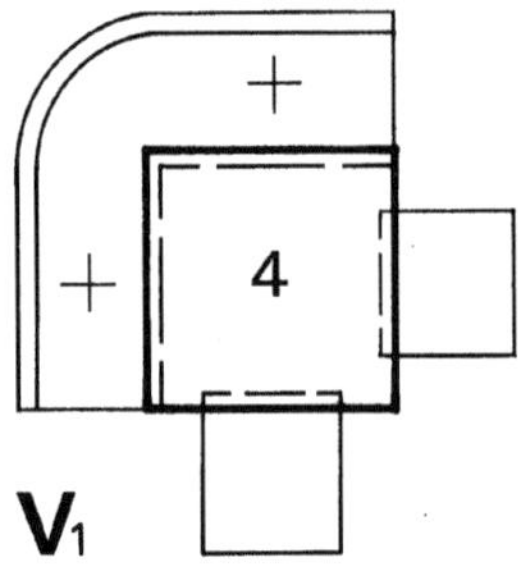

V1

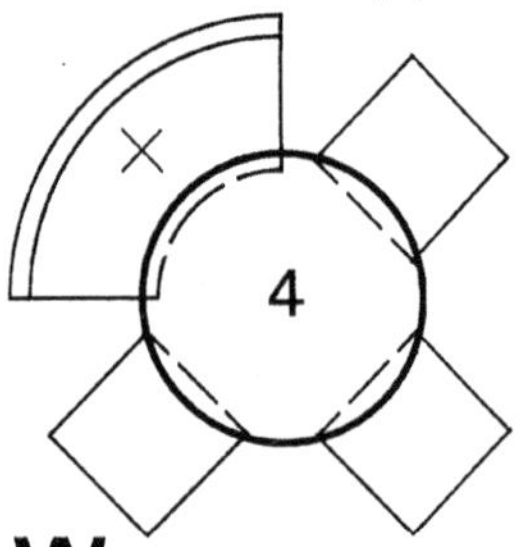

W1

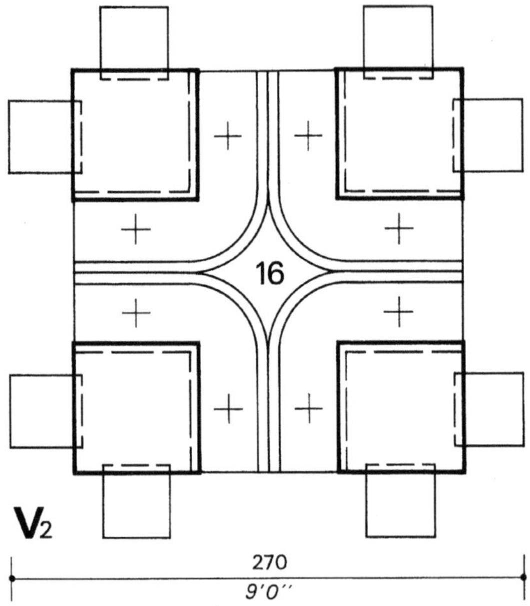

V2

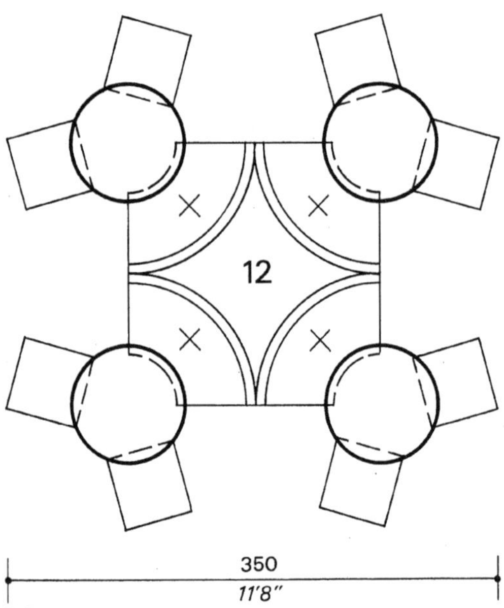

U2

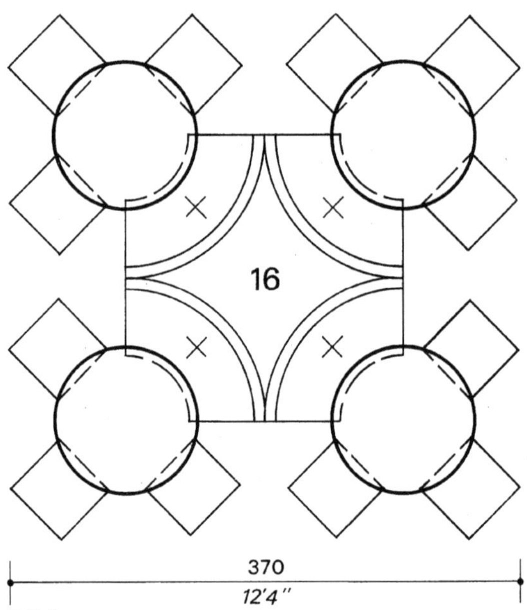

W2

1:50

- Further small arrangements, employing standard 80cm/*32''*, 90cm/*36''* and 100cm/*40''* square tables, and their resulting cluster formations.
- Compared with those on the preceding page, these groupings take perimeter chairs. There is less visual cohesion, but reduced cost outlay on built-in benching.
- U2 is the least spatially efficient.

U1 Corner alcove, with round table, for 3.
U2 Cluster for 12 (derived from U1).
V1 Corner alcove, with square table, for 4.
V2 Cluster for 16 (derived from V1).
W1 Corner alcove, with round table, for 4.
W2 Cluster for 16 (derived from W1).

3.08

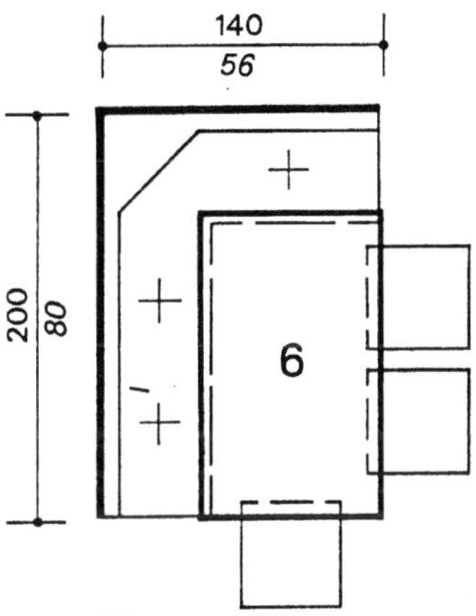

X1

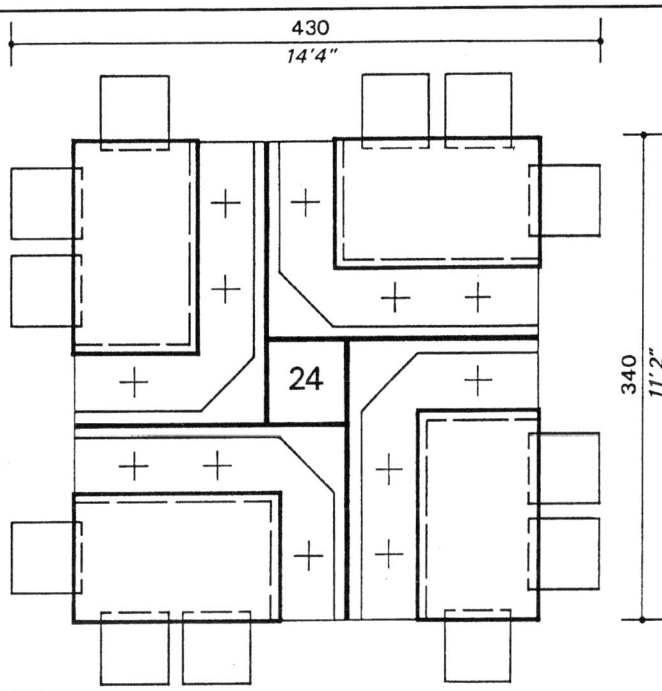

X2

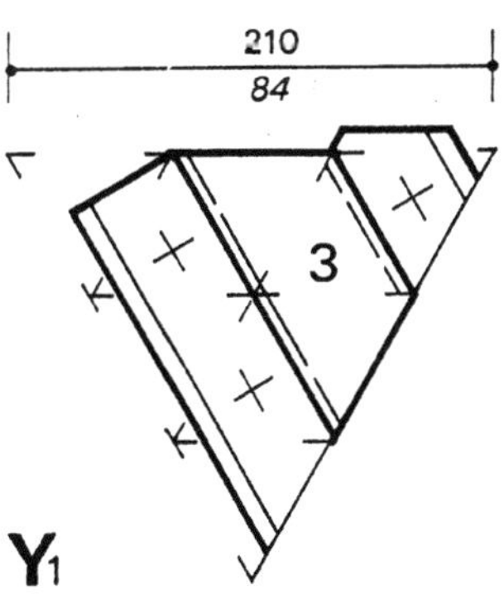

Y1

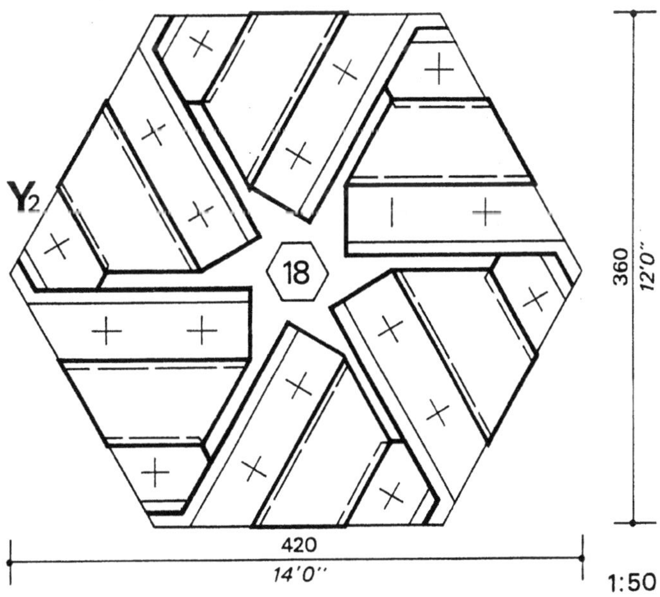

Y2

1:50

- The above examples are based on a 150×90cm/*60×36"* table and a half hexagon.
- X2 is a very practical "windmill" form.
- Y2 is also a "rotating" figure. Something of a *tour de force* to make, unless assembled from six wedge-shaped GRP (fiberglass) mouldings.

X1 Corner alcove, with rectangular table, for 6.
X2 Cluster for 24 (derived from X1).
Y1 Corner alcove, with trapezoid, for 3.
Y2 Cluster for 18 (derived from Y1).

BOOTHS AND BENCHES: summary

1 The two- and four-person booth enjoys an unrivalled popularity on the North America continent, for its implied comfort and intimacy: attributes that may be tuned up or down by varying the basic dimensions, the height of seat backs or intervening screens, and the seat material itself.

2 Wall benches (or banquettes as they are also termed) offer the same comfort but less privacy. When installed in straight runs, as is most common, they also provide the management with the opportunity for varied seating arrangements. Such flexibility is best achieved with small modular tables (introduced in Section 2) and side-seating only.

3 Straight runs may be terminated by returned ends, though care must be taken to ensure that a corner table is long enough to keep neighboring chairs clear of its access path.

4 Curved bays or alcoves – a compromise between booths and straight benching – are visually attractive but spatially less economic. Moreover, any additional loose chairs will protrude into the serving aisle. Part-circular bays, seating 3–6 without an additional chair, are the most successful providing the tables can be extricated for floor cleaning.

5 Small corner alcoves may be combined, back to back, to form free-standing clusters, dominant central features in any room large enough to accommodate them.

The next Section sits people on stools, at serving counters – a fundamentally different design strategy.

COUNTERS AND STOOLS

4

COUNTERS AND STOOLS
Counters and stools: introduction.

4.01

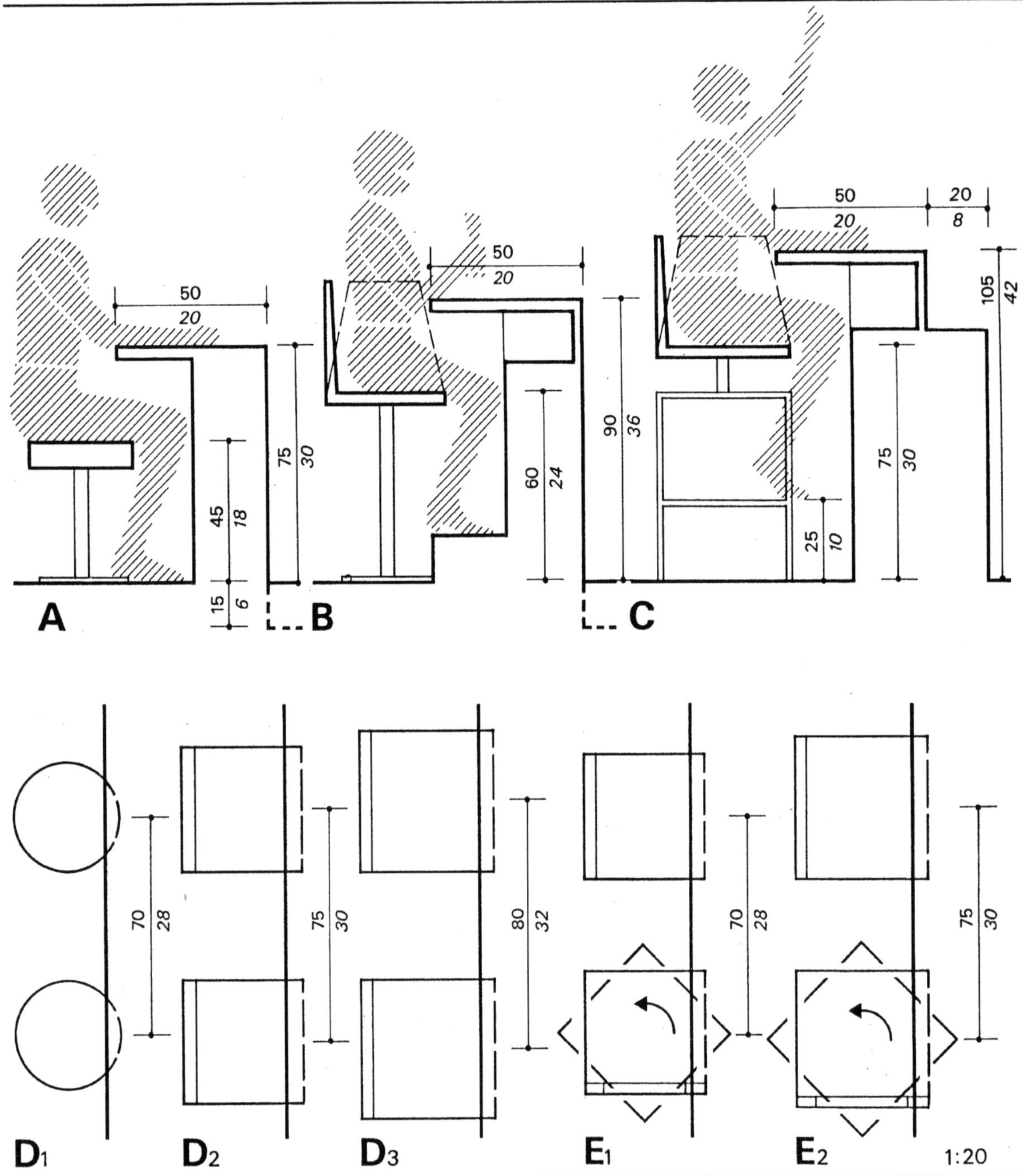

- Typical counter/stool features, illustrated in section and unrelated part plans. Note that types of stool, and counter features other than height, are interchangeable.
- Three *generic* height relationships shown in A, B and C cover the likely range, worldwide:
  LOW: Stool and counter no higher than a normal chair and table, feet on floor.
  MEDIUM: Both somewhat raised, feet on plinth.
  HIGH: Counter now at a convenient serving level, stool (with foot rung) keeping in step.
- It is sometimes possible to achieve a lower service floor (shown in dashed line): not only is the waiter less visually dominant, but his job less back-breaking.

A, B, C Note: (B) the "glove compartment" and (C) the service shelf, where overall width permits.

D1–D3 Three sizes of fixed-seat stool – 35cm/*14″* diameter (backless), 40cm/*16″* and 45cm/*18″* sq. (backed) are spaced to allow 35cm/*14″* customer access. To reduce this clearance, as is sometimes done, makes it a tight squeeze when negotiating high stools.

E1, E2 With swivelling seats, that return by gravity or spring action, customer access widths can safely be lessened by 5cm/*2″*. A tapered back, incidentally, will still allow the counter to overlap the front edge of the seat.

COUNTERS AND STOOLS
Layouts: U-bends – narrow counter, 1 server

4.02

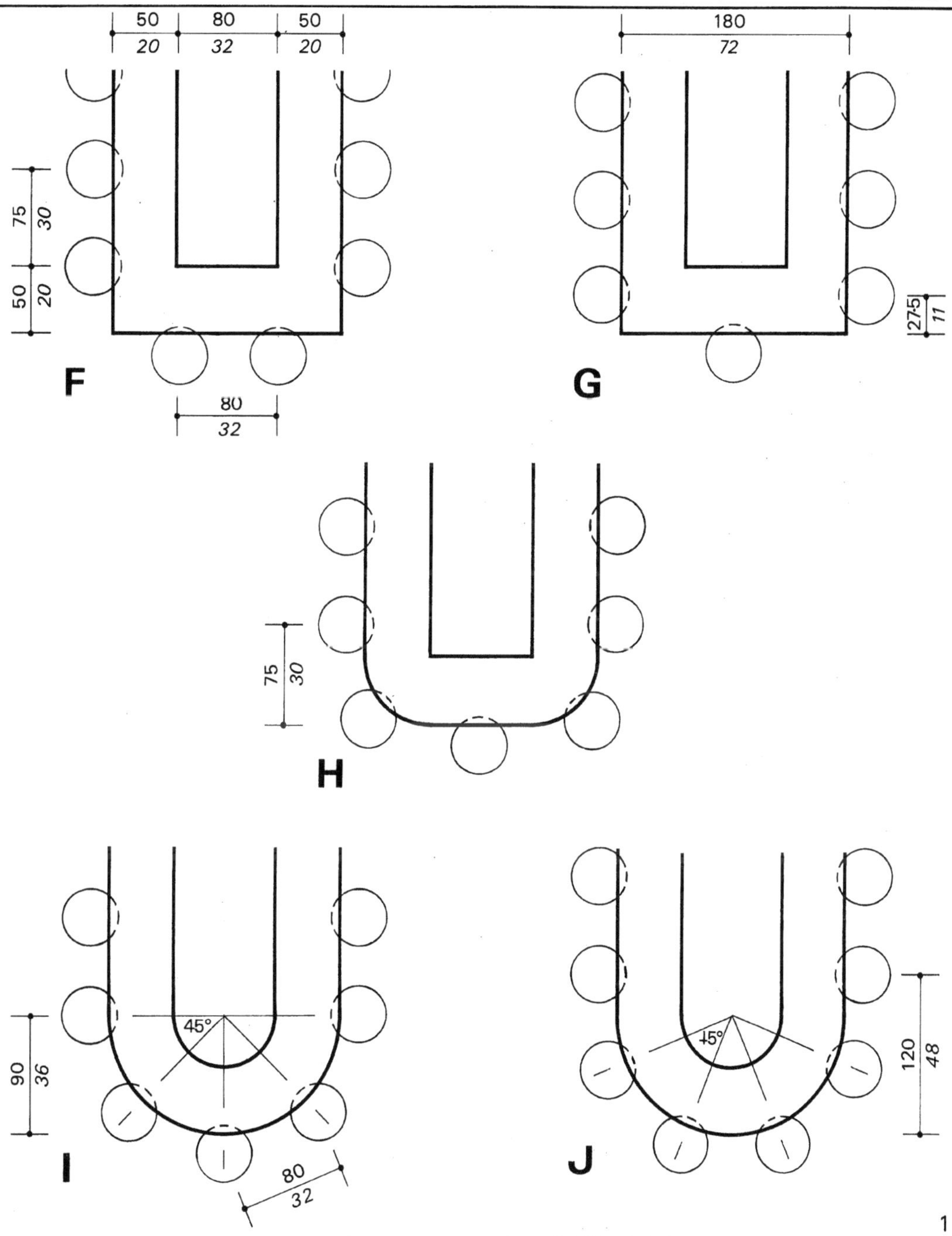

1:50

- Long straight-run counters are simple to design but relatively inefficient in terms of food serving. Many installations consequently employ one or more U-bends, which concentrate a greater number of customers within less distance.
- Here, and on the next three pages, are the basic ways in which such U-bends may be set out. Appropriate stool locations – an important consideration when floor-fixed – are also indicated.
- For purposes of demonstration, all plan diagrams assume a minimum stool spacing of approximately 75cm/*30″*, and the stools themselves are shown round. This first series considers narrow 50cm/*20″* counters, served by one person operating in an adequate width of 80cm/*32″*.
- Apart from G, stool spacings remain visually consistent when turning the corner.

4.03

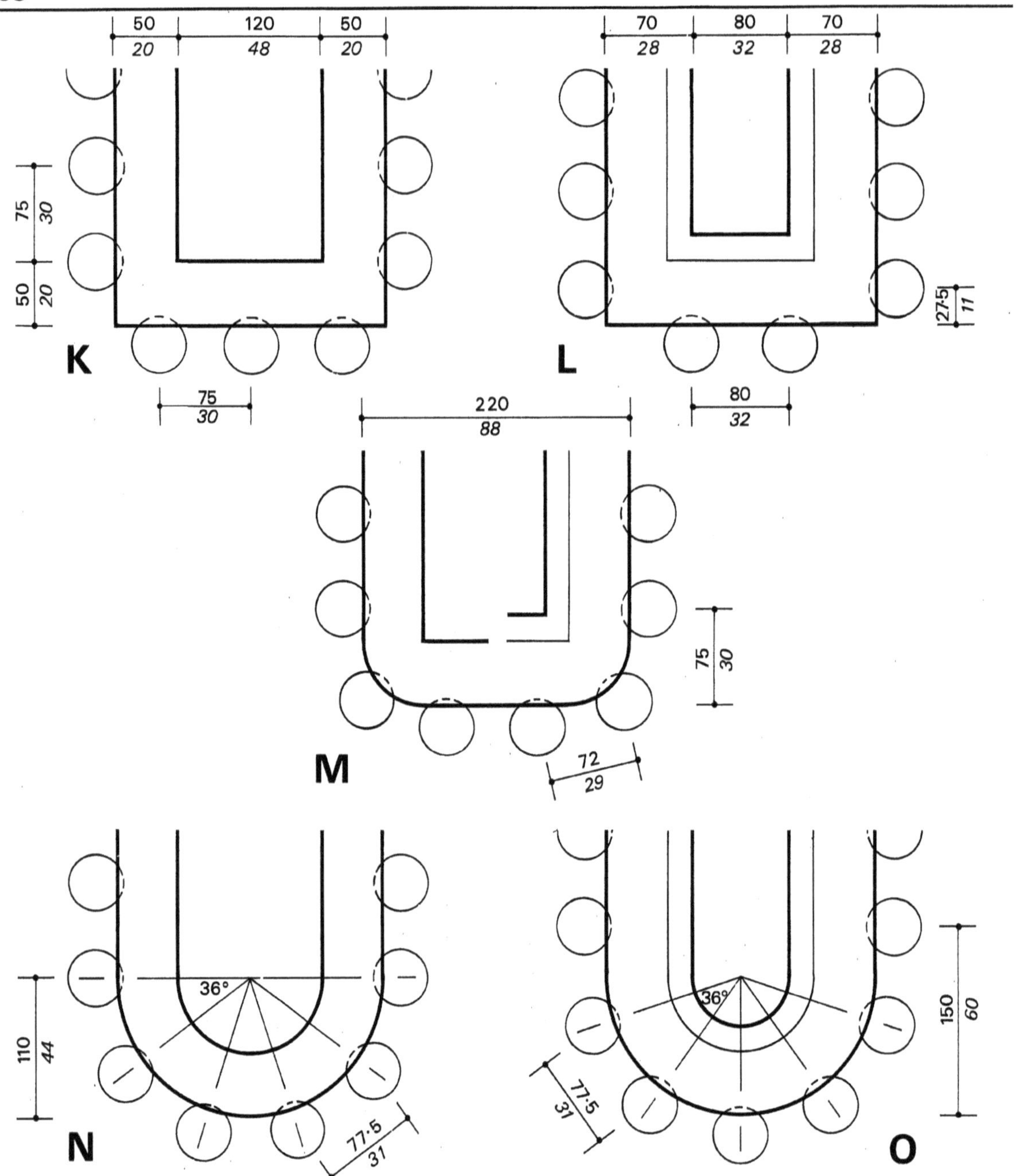

1:50

- Broadening the serving space to accommodate two, or possibly more waiters, adds say 40cm/*16"* – the difference (when totalled) between narrow and wide counter widths. This conveniently permits alternative stool positions to be explored with the same overall layout width of 220cm/*88"*.

K, L Square-cornered counters. If stools are fixed, layout K is the easier to set out and visually better.

M Round-cornered counter, preferably with loose stools.

N, O Looped counters. 36° setting-out works neatly in both cases.

COUNTERS AND STOOLS

Layouts: U-bends – wide counter, 2 servers

4.04

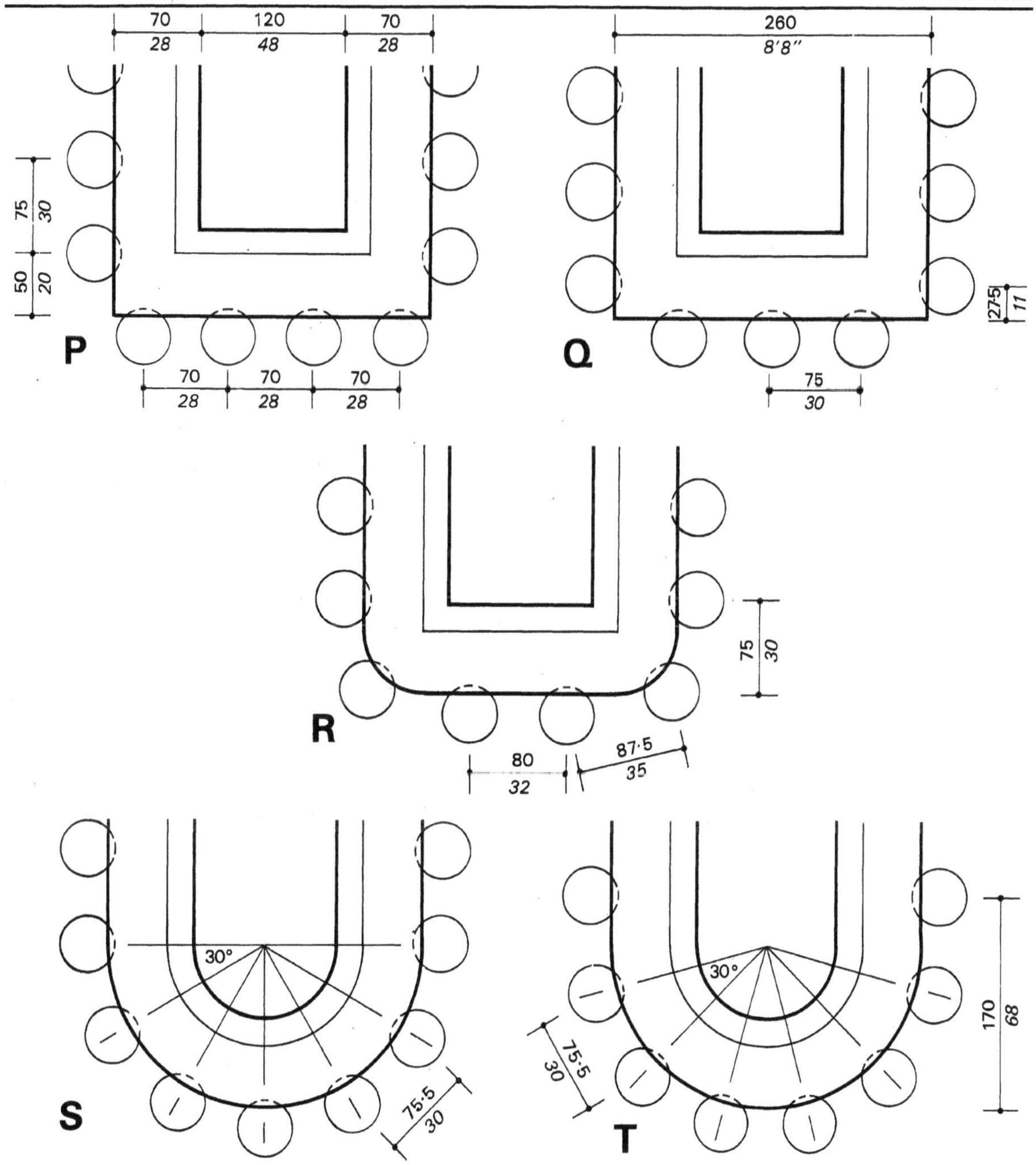

1:50

- Wide U-shape counters, with two servers and an overall width of 260cm/*8′8″*. Arguably, given sufficient room, the most efficient and handsome layouts.

P Note that the four stools, on the flat return end, would need to be slightly closed up.

Q For the same shape, a more satisfactory arrangement.

R Best suited for loose stools.

S, T As on the previous page, full loops are easily set out and attractive to look at.

COUNTERS AND STOOLS

Layouts: U-bends – interior seating

4.05

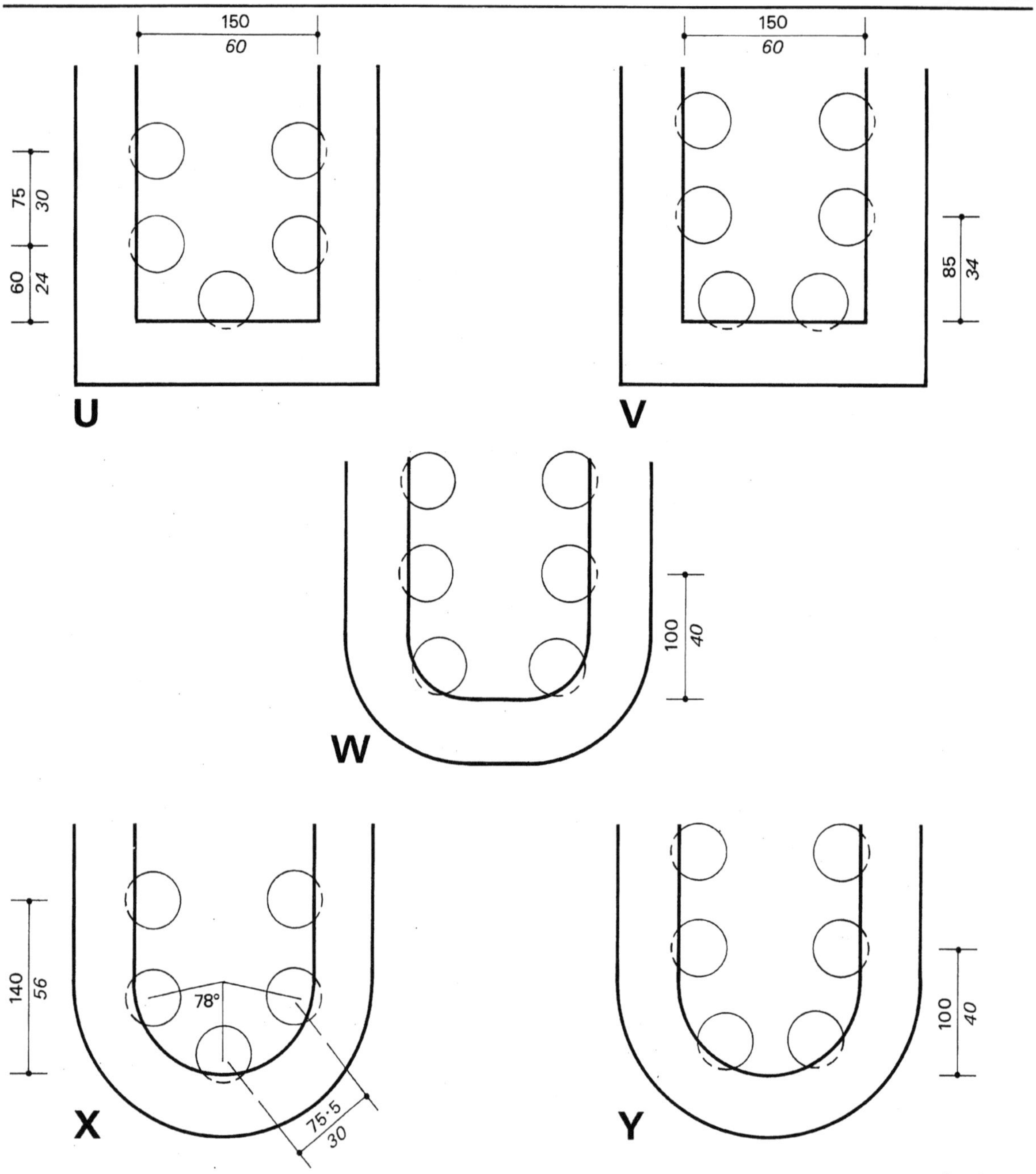

1:50

- Sitting on the *inside* of a counter loop can be called for in re-entrant serpentine layouts, notably in "S" configurations.
- The primary dimension to be established is of course that between opposing counter faces. Around 150cm/*60"* constitutes the minimum width for people to circulate and access their stools.
- The constraints imposed by tight cornering, and the resultant wastefulness of personal counter surface, are clearly evident in these examples.

U, X Afford the best visual continuity.

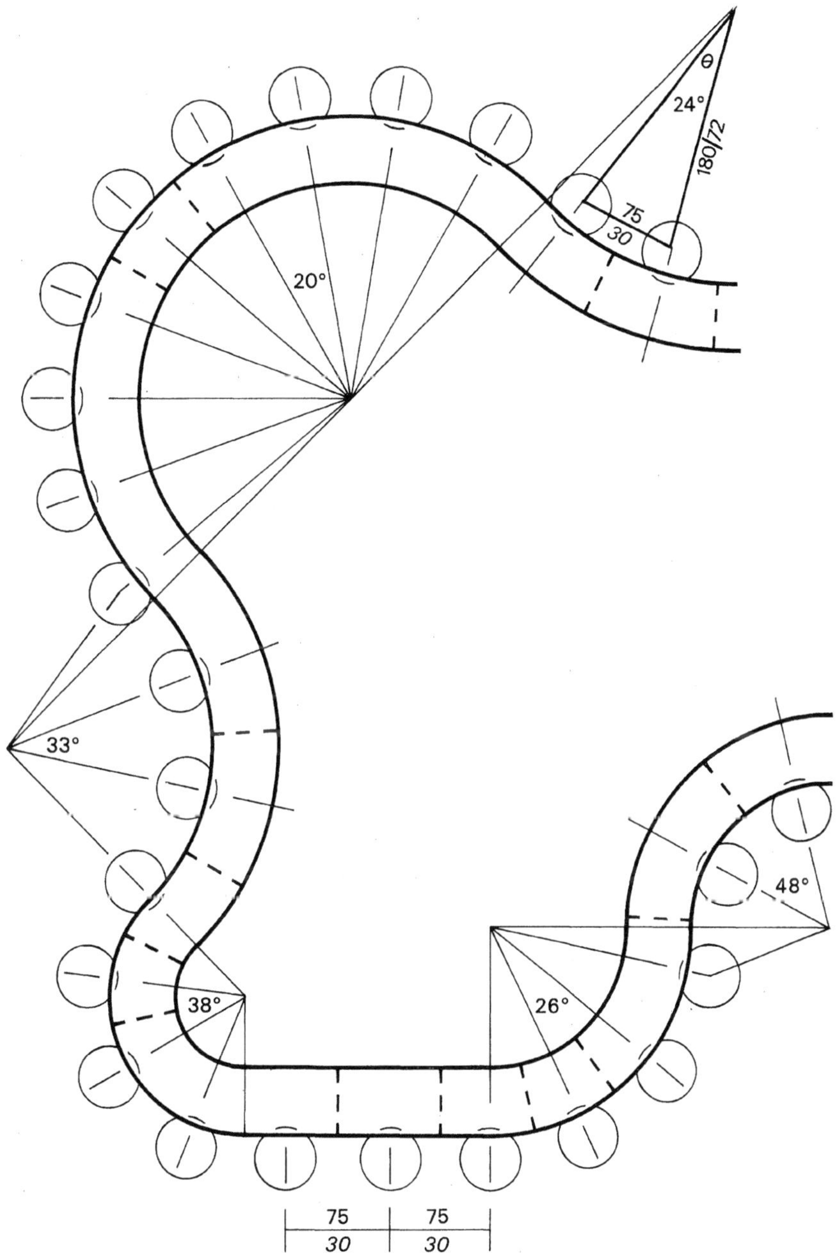

1:50

- As has been seen, a great advantage of counter seating is the ability to create curves. Free-form serpentine layouts, albeit space-hungry, are theoretically most tempting but would make severe demands on the skills of designers, fabricators and installers.
- Note also how sinuosity confers considerable variety of individual counter space, when compared with the (already over-generous) straight-run allocation.
- This hypothetical example, which shows all construction lines, highlights the geometric complexities of the setting-out: particularly onerous if stools are to be *fixed* and at the same centers.
- Such stool spacing – e.g. top right – may be calculated by the formula

$$\mathbf{C} = 2\mathrm{R}.\ \sin\frac{\theta}{2}$$

where
C = *required spacing (chord)*
R = *radius of stool centers*
θ = *subtending angle*
(See APP. 3 GEOMETRIC PROPERTIES for a selection of the more common subtending angles and radius/chord ratios.)

COUNTERS AND STOOLS: summary

1. Lunch counters (by any other name and wherever situated) are arguably the quick turnover forerunners of the cafeteria, but with an essential difference: customers are seated by the food and drink source.
2. Operational efficiency, in preparing and serving food, requires that the staff restrict their movements to the minimum. Counters, therefore, ideally are high and wrap-around where space allows. This makes for U-bend loops.
3. Since stools, even when not fixed to the floor, can not easily be pulled up like chairs, sufficient intervening access space must be allowed. Consequently – though swivelling stools can somewhat reduce this clearance – personal place width tends to be more generous than that at a table. This coincides with the fact that many diners will be on their own.
4. The higher and wider the overall counter the better. The waiter becomes less visibly dominant, and useful shelves can be incorporated for customers' accessories and for serving.
5. Free-form serpentine layouts are possible, but entail manufacturing and installation problems coupled with circulation constraints for any re-entrant curves.

The next Section examines tables, chairs and bench seating in their larger context.

MODULES AND PLANS

5

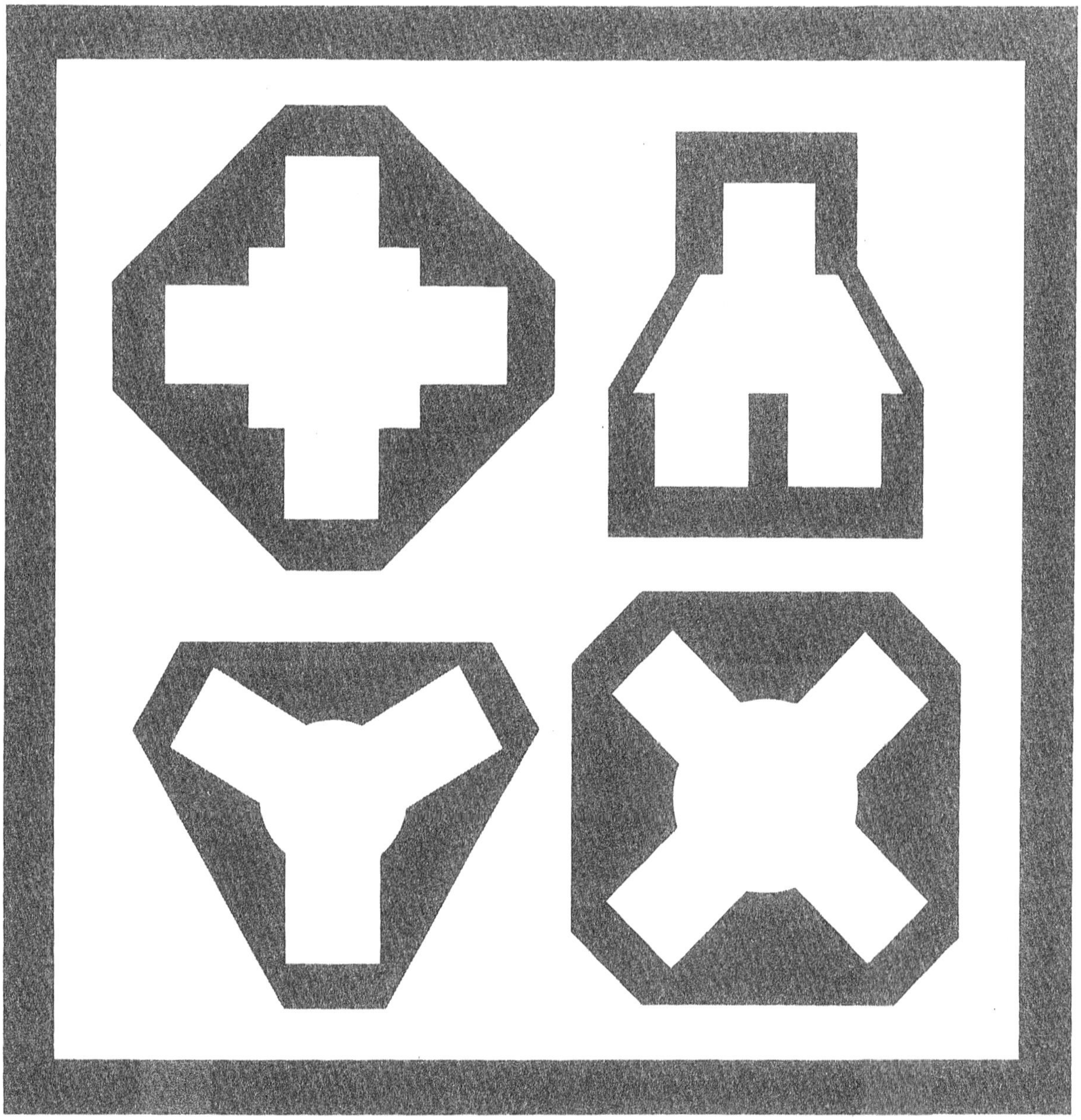

MODULES AND PLANS
Modules: introduction

5.01

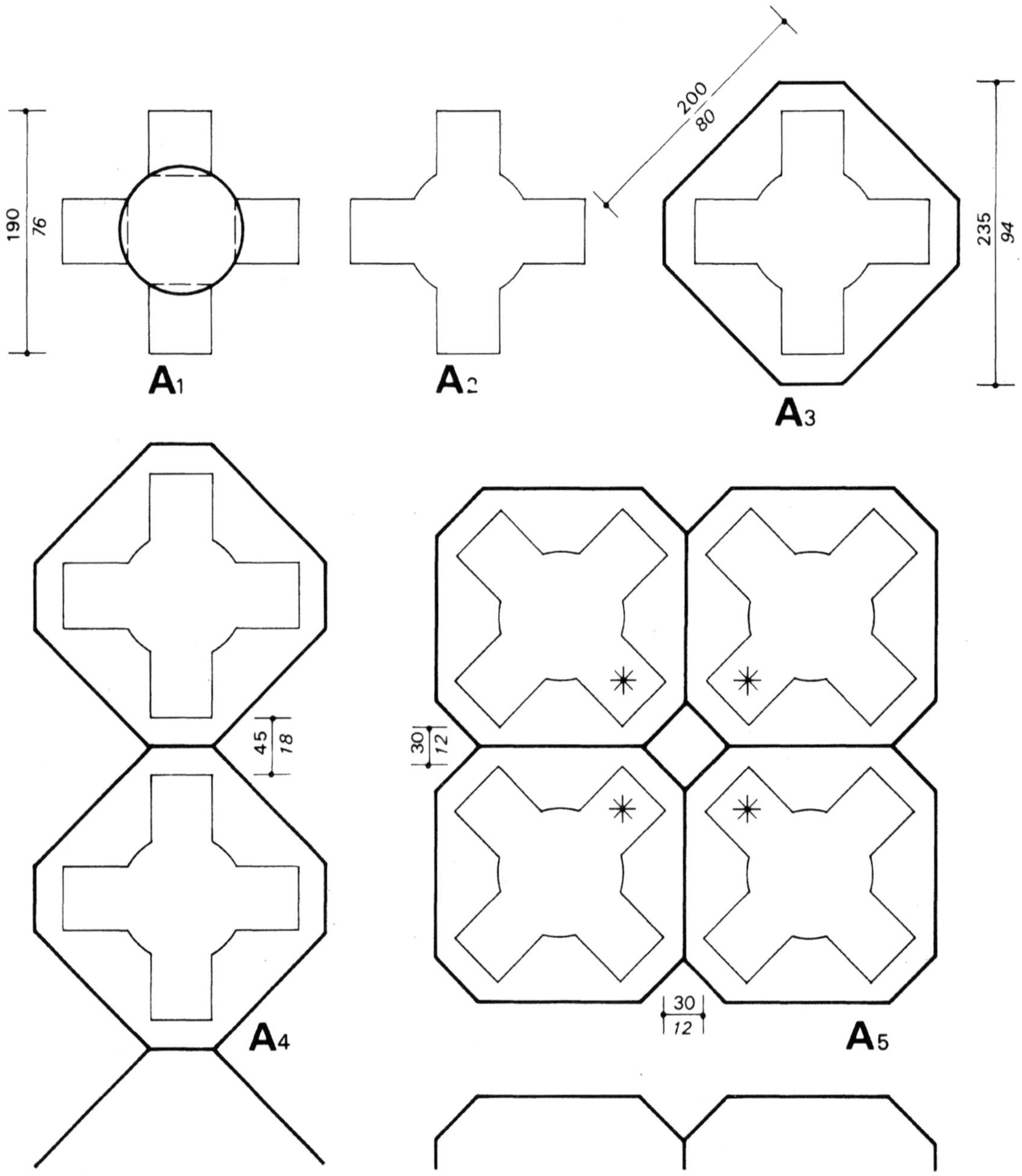

1:50

- The next logical and essential step, after establishing suitable individual table-and-seat shapes and sizes, is to consider how these affect and combine with one another in the broader spatial context of room plans.

A1 Round table and four chairs – an individual ***assembly***.

A2 The assembly's ***footprint***.

A3 The footprint's immediate territorial ***envelope*** (plan module): "immediate" in the sense that it includes a small (22.5cm/*9"*) seat-accessing allowance but does **not** – as some space plans do – incorporate a half share of an adjoining, variable circulation aisle.

A4 When doubled, between opposed chair backs, the seat access allowance obligingly provides a reasonable *local* customer clearance of 45cm/*18"*.

A5 In a diagonally-oriented group of four tables, this clearance reduces to a scarcely negotiable 30cm/*12"*. Getting to the innermost seats (starred) is perhaps best achieved by operating the "first in, last-out" principle. Such a group compensates, however, by being more compact than the rectangular version A4, which donates excessive area to the aisles.

MODULES AND PLANS

Modules: side-seating rectangular tables, seating 1–8

5.02

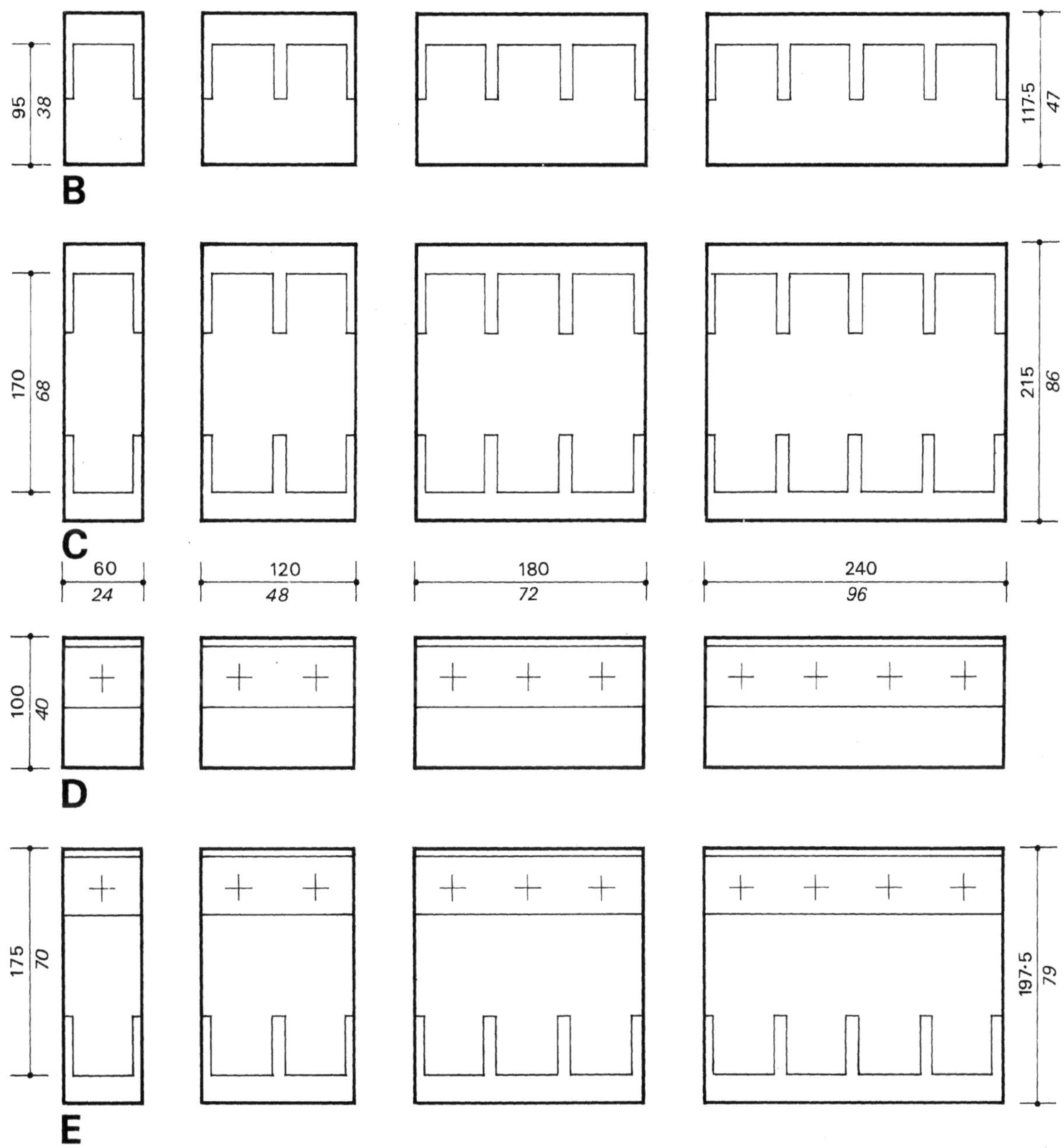

1:50

- ***"Envelope"*** modules for a range of side-seating rectangular tables, free-standing (B, C) and bench-related (D, E) sized in accordance with 2.01.

B, D Unilateral tables, width 50cm/*20"*.
C, E Bilateral tables, width 80cm/*32"*.

MODULES AND PLANS
Modules: side- and end-seating rectangular tables, seating 4–8

5.03

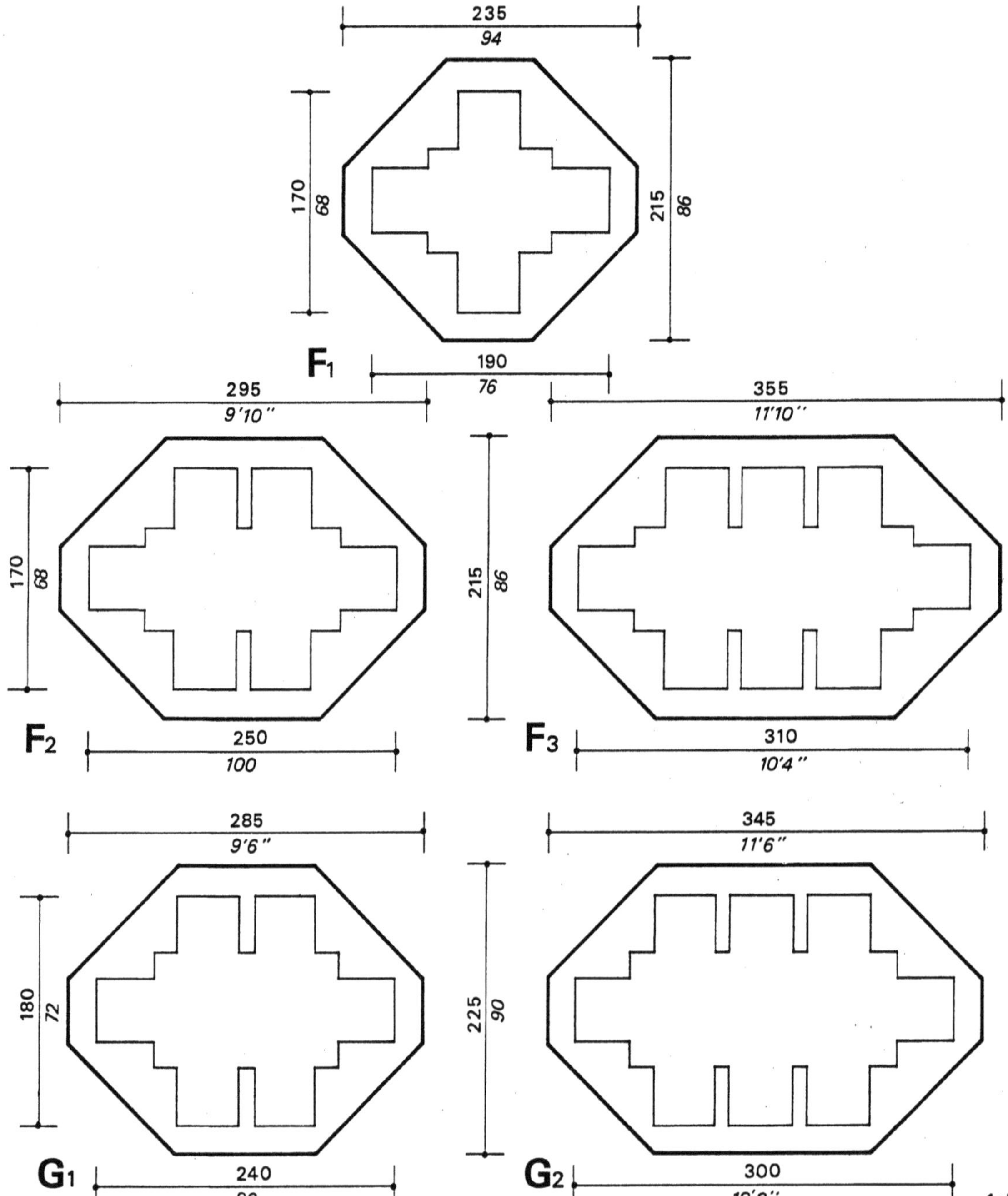

1:50

- ***Envelope*** modules for a range of perimeter-seating rectangular tables, sized in accordance with 2.02.

F1–F3 Table width 80cm/*32"*.
G1, G2 Table width 90cm/*36"*.

Modules: round and square tables, seating 2–10

5.04

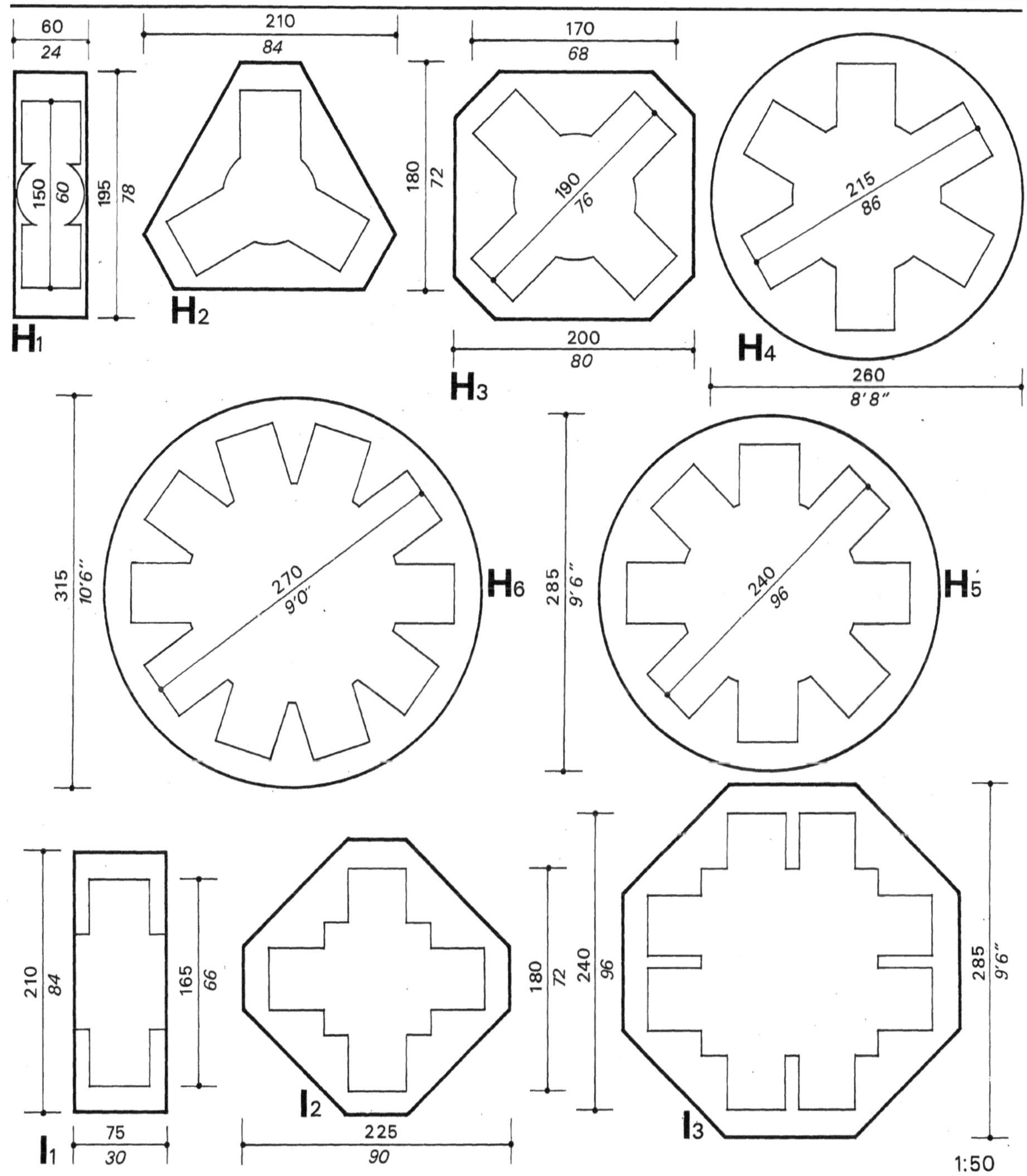

- ***Envelope*** modules for a range of round and square tables, sized as established in 2.01, 2.02, 2.03, 2.04, 2.11 and 2.12.

Round table *diameters* (persons):

H1 60cm/*24″* (2).
H2 80cm/*32″* (3).
H3 100cm/*40″* (4).
H4 125cm/*50″* (6).
H5 150cm/*60″* (8).
H6 180cm/*72″* (10).

Square table *sides* (persons):

I1 75cm/*30″* (2).
I2 90cm/*36″* (4).
I3 150cm/*60″* (8).

Modules: oval tables, seating 3–8

5.05

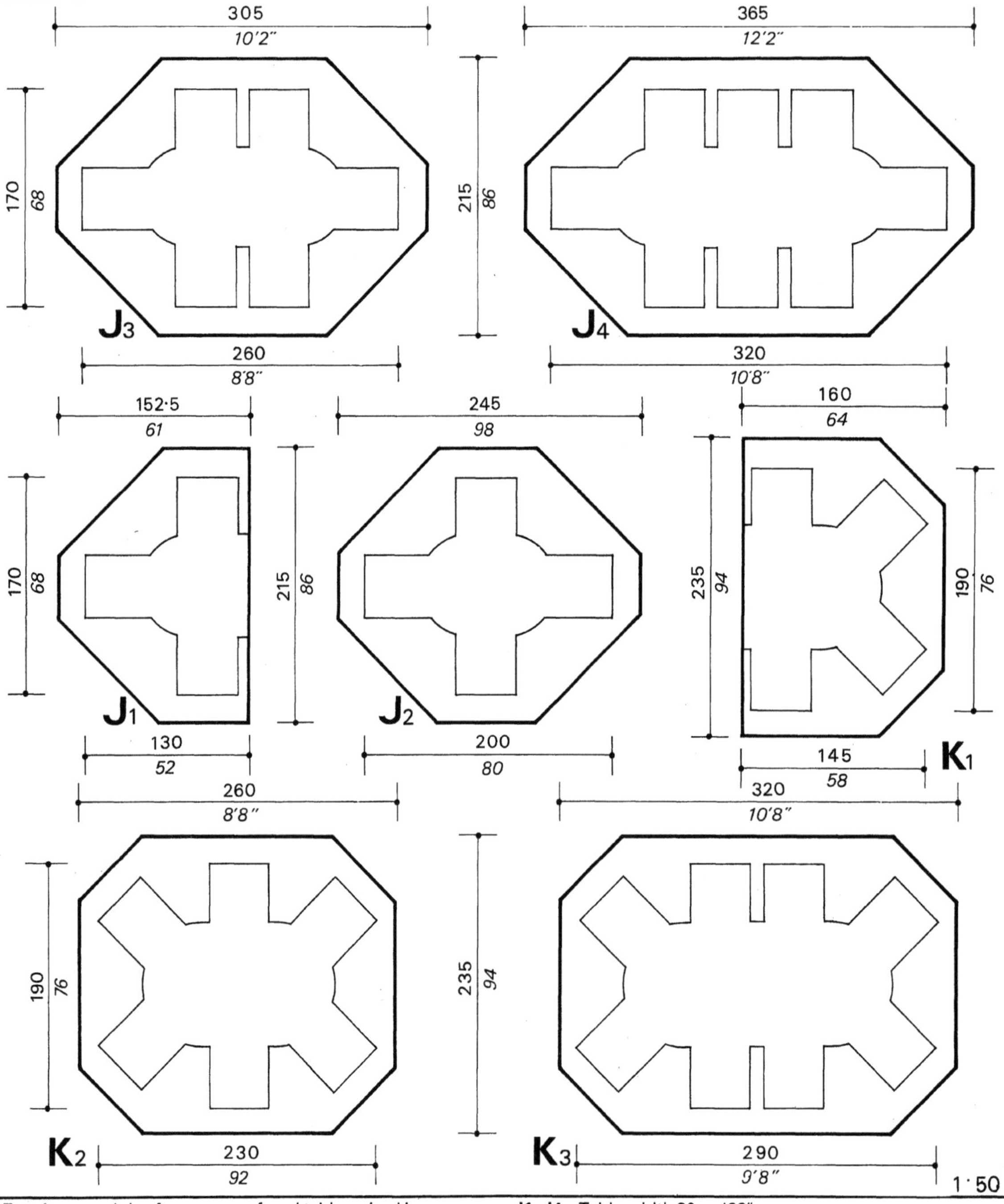

- ***Envelope*** modules for a range of oval tables, sized in accordance with 2.05.

J1–J4 Table width 80cm/*32″*.
K1–K3 Table width 100cm/*40″*.

Modules: triangular-based tables, seating 2–8

5.06

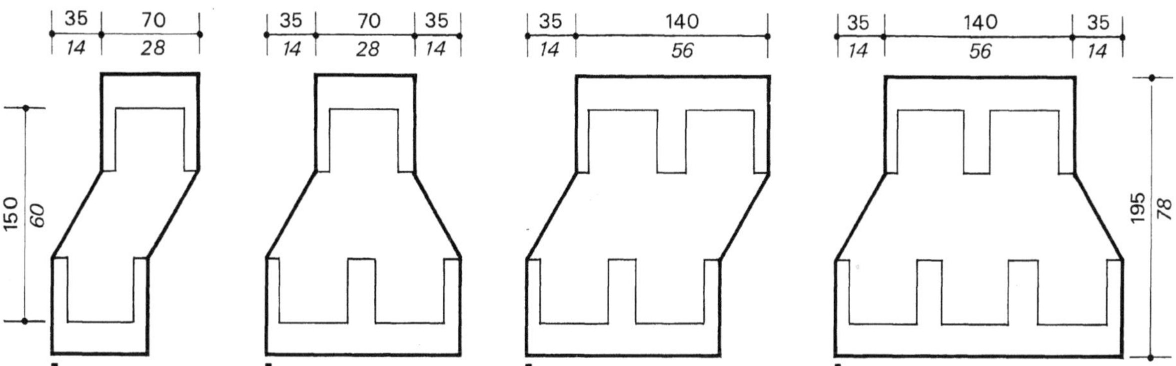

L1 L2 L3 L4

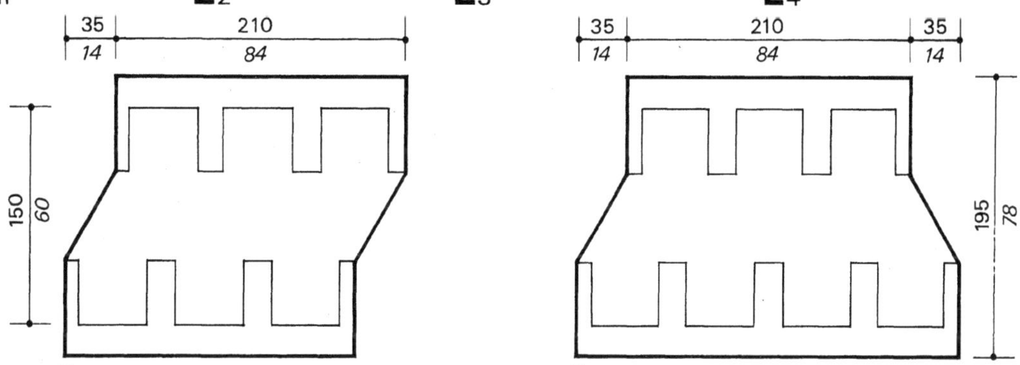

L5 L6

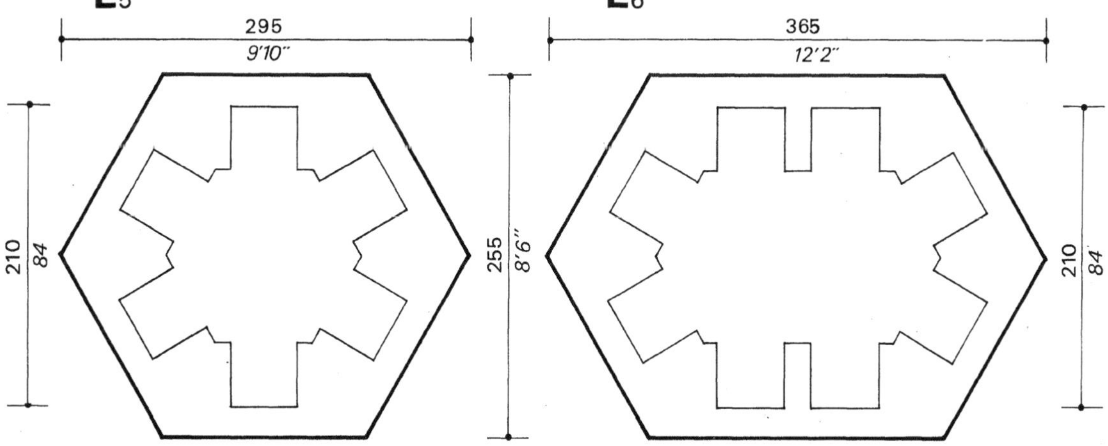

M1 M2

1:50

- ***Envelope*** modules for a range of tables based on the equilateral triangle, sized in accordance with 2.10. All table widths are determined by 70cm/*28″* triangle sides.

L1-L6 Table width fractionally over 60cm/*24″*.
M1, M2 Table width fractionally over 120cm/*48″*.

Modules: comparative densities – tables seating 2 and 4

5.07

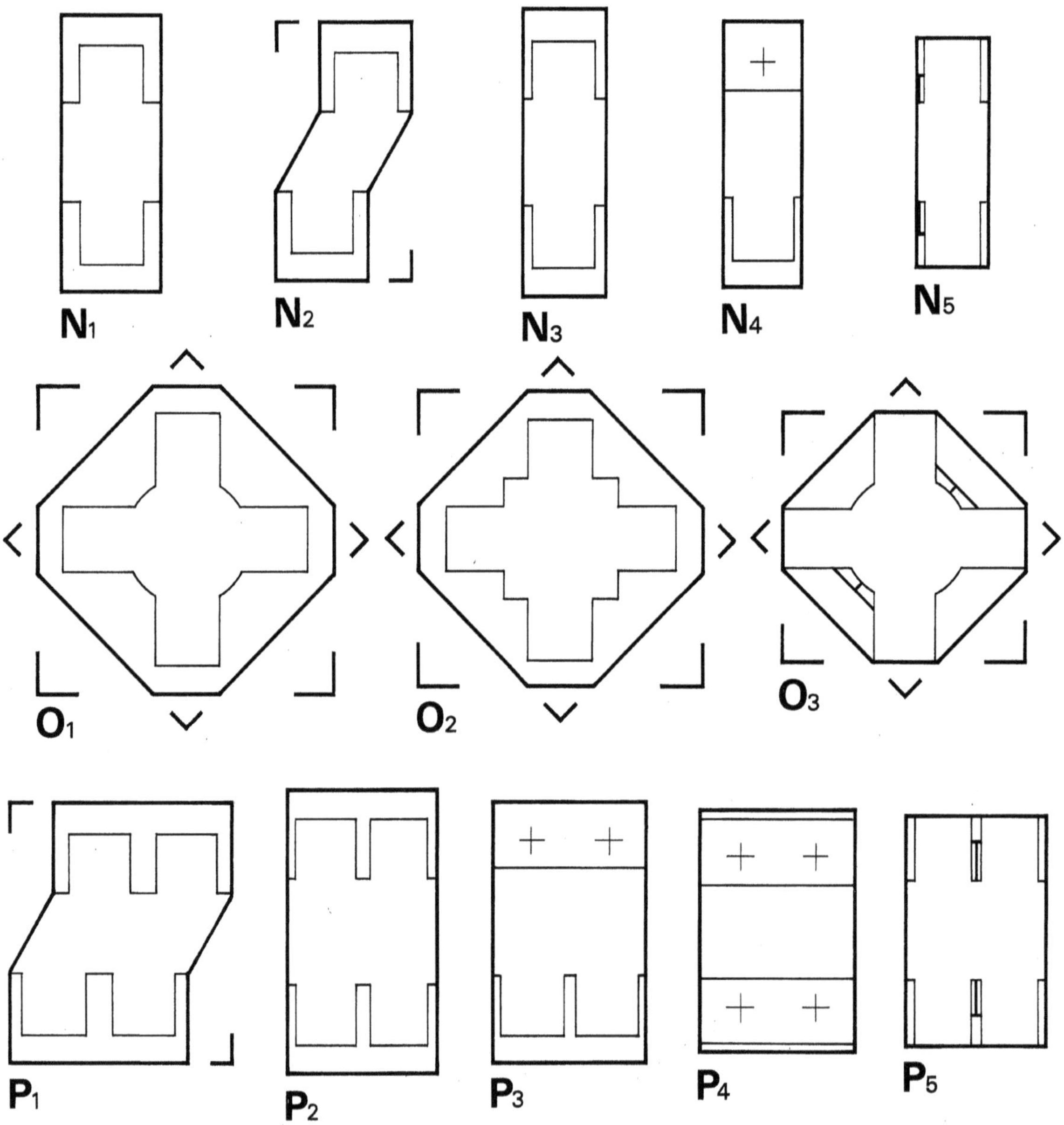

1:50

Guidelines
For *average* tables, allow module areas as follows:
Conventional table for 2 – 1.4m²/*15ft²*
Table-seat unit for 2 – 0.9m²/*10ft²*
Conventional table for 4 – 3.0m²/*32ft²*
Table-seat unit for 4 – 2.2m²/*24ft²*

	2-person table category	table size cm/*in*	module area m²/*ft²*
N1	Free-standing	75/*30* square	1.6/*17*
N2	Free-standing	70×70/*28×28*	1.4/*15*
N3	Free-standing	60×80/*24×32*	1.3/*14*
N4	Alongside bench	60×80/*24×32*	1.2/*13*
N5	Table/seat unit	53×75/*21×30*	0.9/*10*

	4-person table category	table size cm/*in*	module area m²/*ft²*
O1	Free-standing	100/*40* diam.	3.8/*41*
O2	Free-standing	90/*36* square	3.5/*38*
O3	Table/seat unit	95/*38* diam.	2.6/*28*
P1	Free-standing	70×140/*28×56*	2.7/*29*
P2	Free-standing	80×120/*32×48*	2.6/*28*
P3	Alongside bench	80×120/*32×48*	2.4/*26*
P4	Booth	70×120/*28×48*	2.2/*24*
P5	Table-seat unit	75×105/*30×42*	1.8/*19*

MODULES AND PLANS

Modules: comparative densities – tables seating 6

5.08

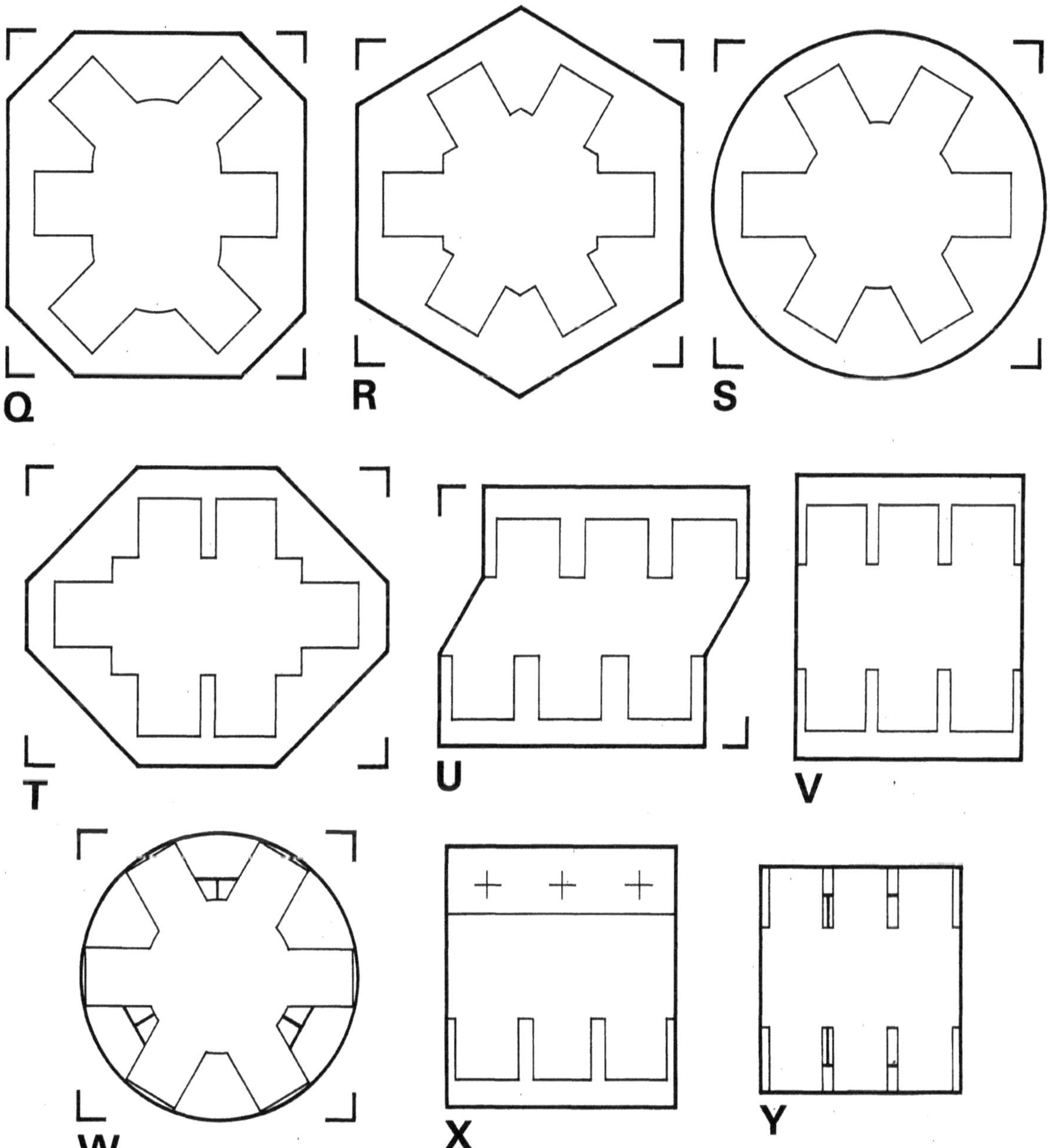

1:50

Guidelines

For *average* tables, allow module areas as follows:
Conventional table for 6 – 4.6m²/*49ft²*
Table-seat unit for 6 – 3.3m²/*35ft²*

	6-person table category	table size cm/*in*	module area m²/*ft²*
Q	Free-standing	160×100/*64×40*	5.6/*60*
R	Free-standing	140×121/*56×48*	5.5/*59*
S	Free-standing	125/*50* diam.	5.3/*57*
T	Free-standing	90×150/*36×60*	4.9/*53*
U	Free-standing	70×210/*28×84*	4.0/*43*
V	Free-standing	80×180/*32×72*	3.9/*42*
W	Table/seat unit	115/*46* diam.	3.8/*41*
X	Alongside bench	80×180/*32×72*	3.5/*38*
Y	Table/seat unit	75×158/*30×63*	2.7/*29*

Note: For initial planning purposes it is the larger *room density* that matters most. This can only be approximated by taking into account important variables such as juxtaposition, groupings, setting-out axes, circulation routes and available width-factors investigated in the remaining pages of this Section.

MODULES AND PLANS

Access: customer to seat, server to tableware

5.09

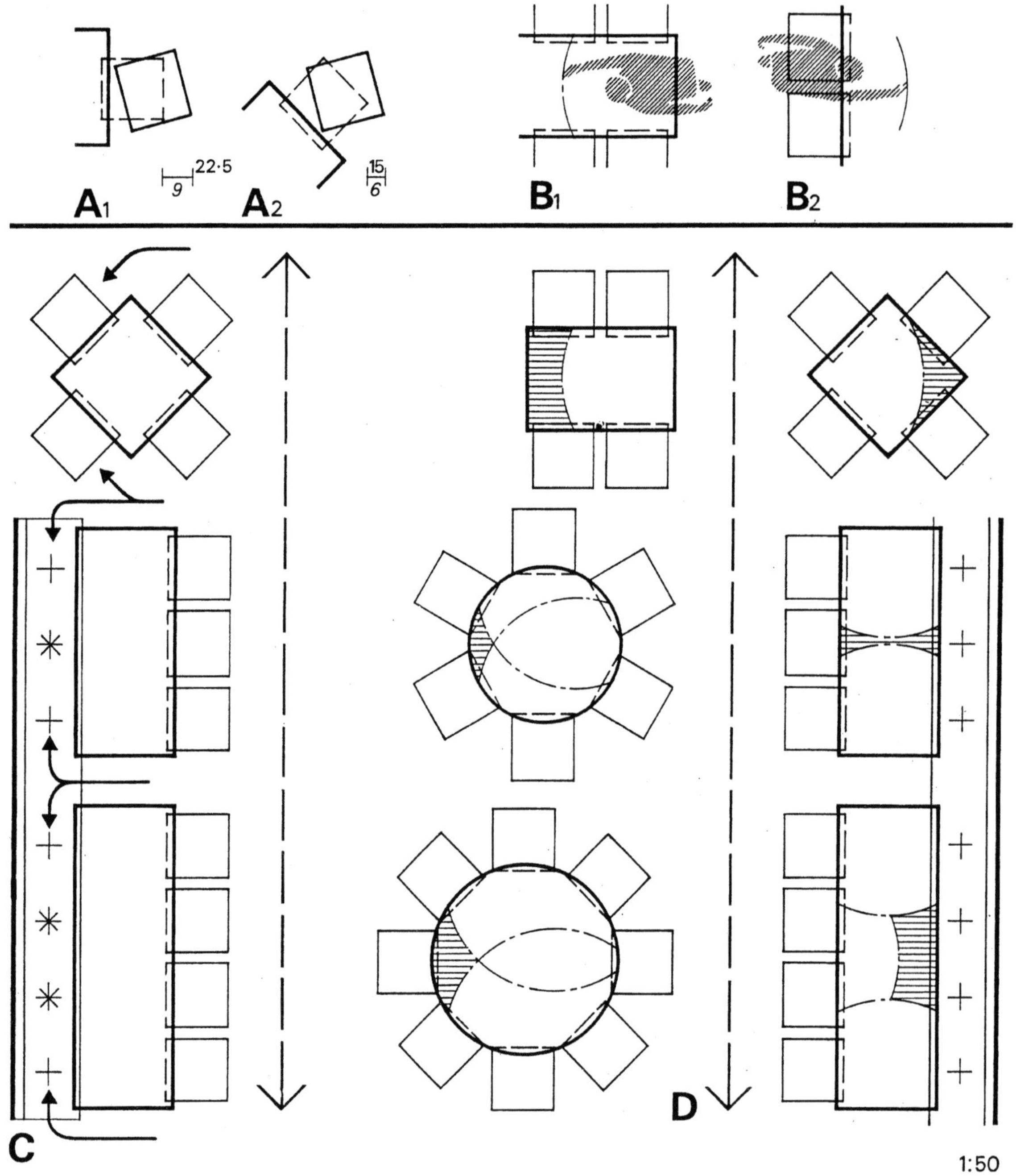

- **Access** – to seats, tableware and dining surfaces, is a fundamental consideration. Various aspects, some touched on earlier, are examined here:

A1, A2 Getting in and out of a chair: allow an average 20cm/*8"*.

B1 Waiter's reach, leaning well-forward from the table end and unimpeded by diners: assume approx. 90cm/*36"* from the table edge.

B2 Waiter's reach, serving between occupied seats and partially impeded: assume say 50cm/*20"* from the table edge. Most diners will shift over as a normal reflex.

C Bench seat access, usually from either side of the table: allow 35cm/*14"* minimum clearance and anticipate "first in, last out" behavior for central positions (starred).

D Table areas beyond a waiter's reasonable grip-reach (see comments, B1, B2) are shown shaded in this composite layout of typical tables served from a central aisle. Two points are worth noting:

- When rectangular tables are pushed together for parties of 8+, the server may require some "give and take" from those sitting on the middle bench seats.
- Circular tables seating 8+ ideally require all-round serving.

MODULES AND PLANS

Access: corner layout constraints

5.10

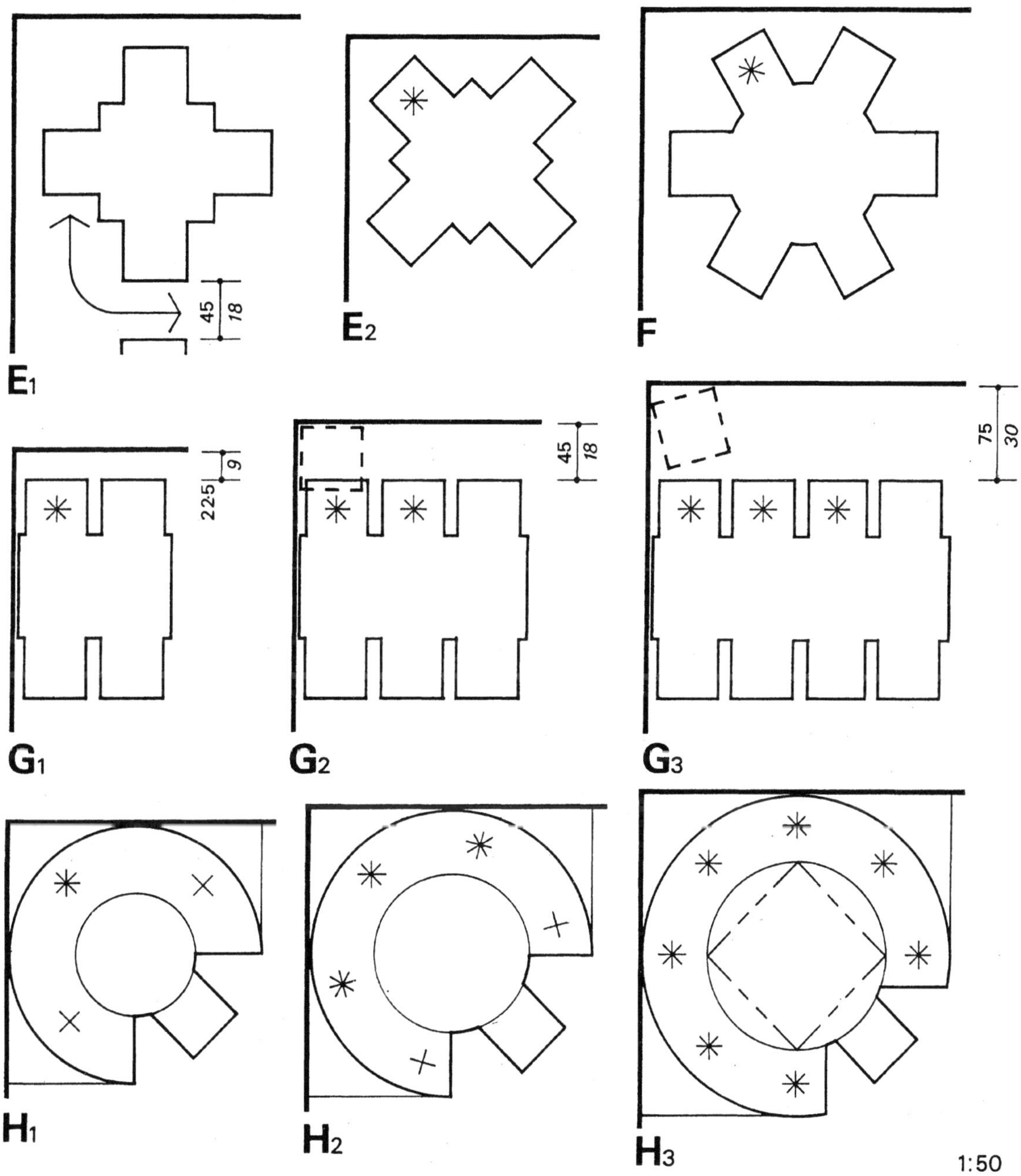

• Economic space utilization suggests that positioning of corner tables and chairs be determined, as elsewhere, by *envelope* modules. This allows for sitting down and getting up, but may well restrict access to furthermost seats.

E1 4-seater square table placed perpendicularly: no problem.

E2, F, G1 The same table placed diagonally, a round 6-seater and rectangular 4-seater: in each case the corner seat is bottled up. However, if the occupant should need to join or leave the table during the meal, it is quite reasonable to expect his neighbor to let him through.

G2, G3 With three or more adjacent diners, the seat access problem increases: progressively wider margins (45cm/*18″* and 75cm/*30″*) are consequently recommended. The greater of the two clearances *will also permit table service.*

H1–H3 Built-in corner bays are environmentally attractive, and to sit and slide round is an accepted sequential procedure, though inconvenient if there is frequent movement among diners.

MODULES AND PLANS

Access: bench seating and aisle width considerations

5.11

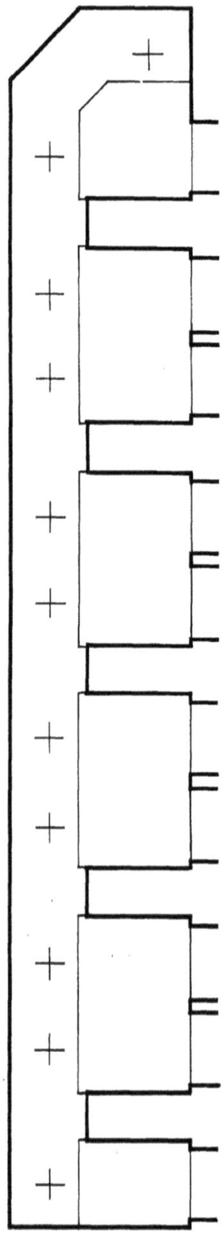

I_1

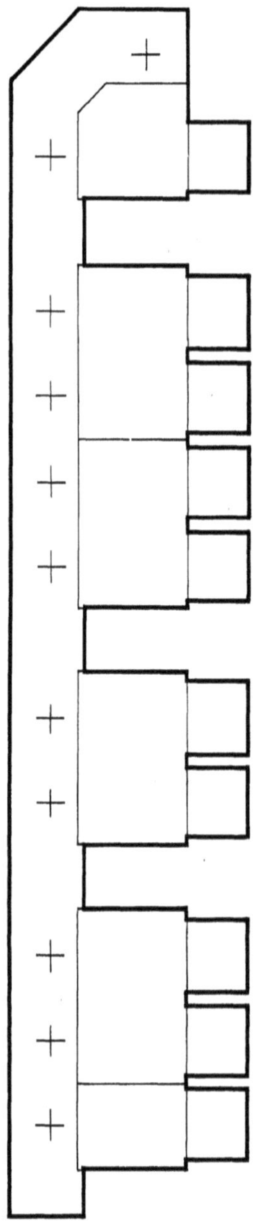

I_2

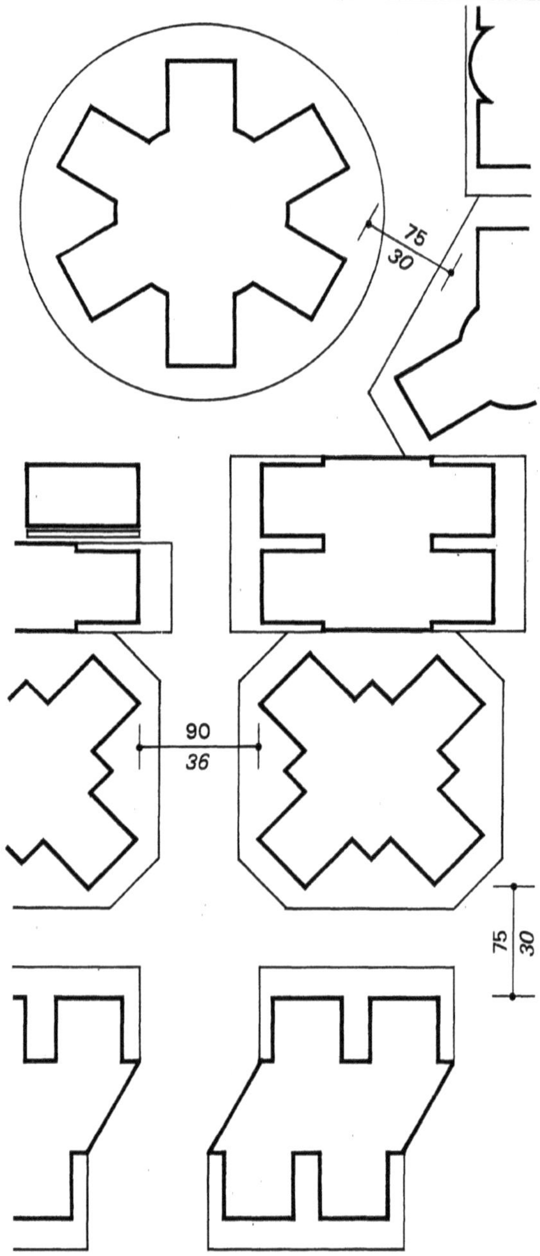

J

1:50

- Three pertinent observations concerning flexibility of access may be made in regard to bench seating and circulation aisles:
- Bench seating: when small tables are joined to form longer ones, the access gaps reduce in number and fortuitously widen, but the resulting redistribution of bench seat positions also cuts down local wear and tear of the upholstery.
- Aisle direction: need not be straight. Angling an aisle, relative to the major axes, may respond better to particular table configurations (and vice versa).
- Aisle width: need not be constant. Nor can it be, unless flanked on both sides by closely aligned rectangular tables. It is the overall tabling pattern that predominates and establishes the margins.
- Moreover, since chair access is an intermittent activity, it is the succession of furniture edges – the *footprint* outlines – that form an aisle's real boundary. To the extent that this irregular boundary departs from the theoretical one, small or large passing places will be created. Providing an appropriate minimum width is preserved, e.g. as an escape route or for a piece of equipment, the designer can take these bonuses into account.

I1, I2 Typical bench run, before and after table redistribution.

J Composite layout, illustrating changes of aisle angle and width.

5.12

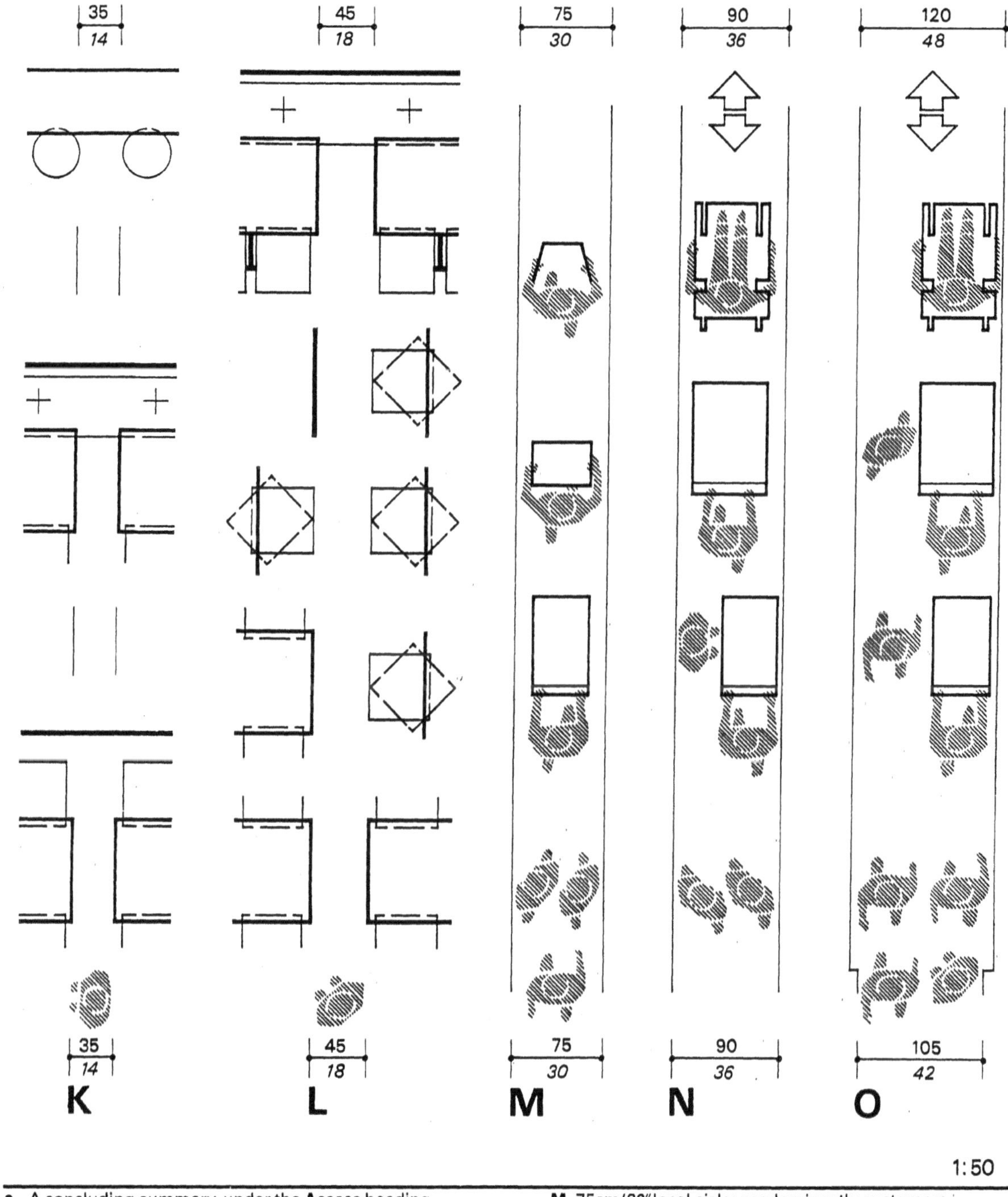

- A concluding summary, under the **Access** heading, of recommended local clearances and aisle widths. Though finer distinctions could be made, six easily remembered dimensions will adequately cover all normal eventualities:

K 35cm/*14″* personal seat access between table ends and counter stools.

L 45cm/*18″* personal seat access between table/seat units, and short-distance customer access between chairs, tables and walls.

M 75cm/*30″* local aisle: predominantly customers, occasional service.

N 90cm/*36″* local aisle: customers and regular service.

O 105–120cm/*42″–48″* main aisle: all-purpose, including escape route.

- The scaled human figures – to be seen in a wheelchair, with a walking frame or tray, and pushing food and clearing trolleys – assist the reader's perception of the various widths. Note the progression from crabbing sideways through the narrowest gap to striding two abreast.

5.13

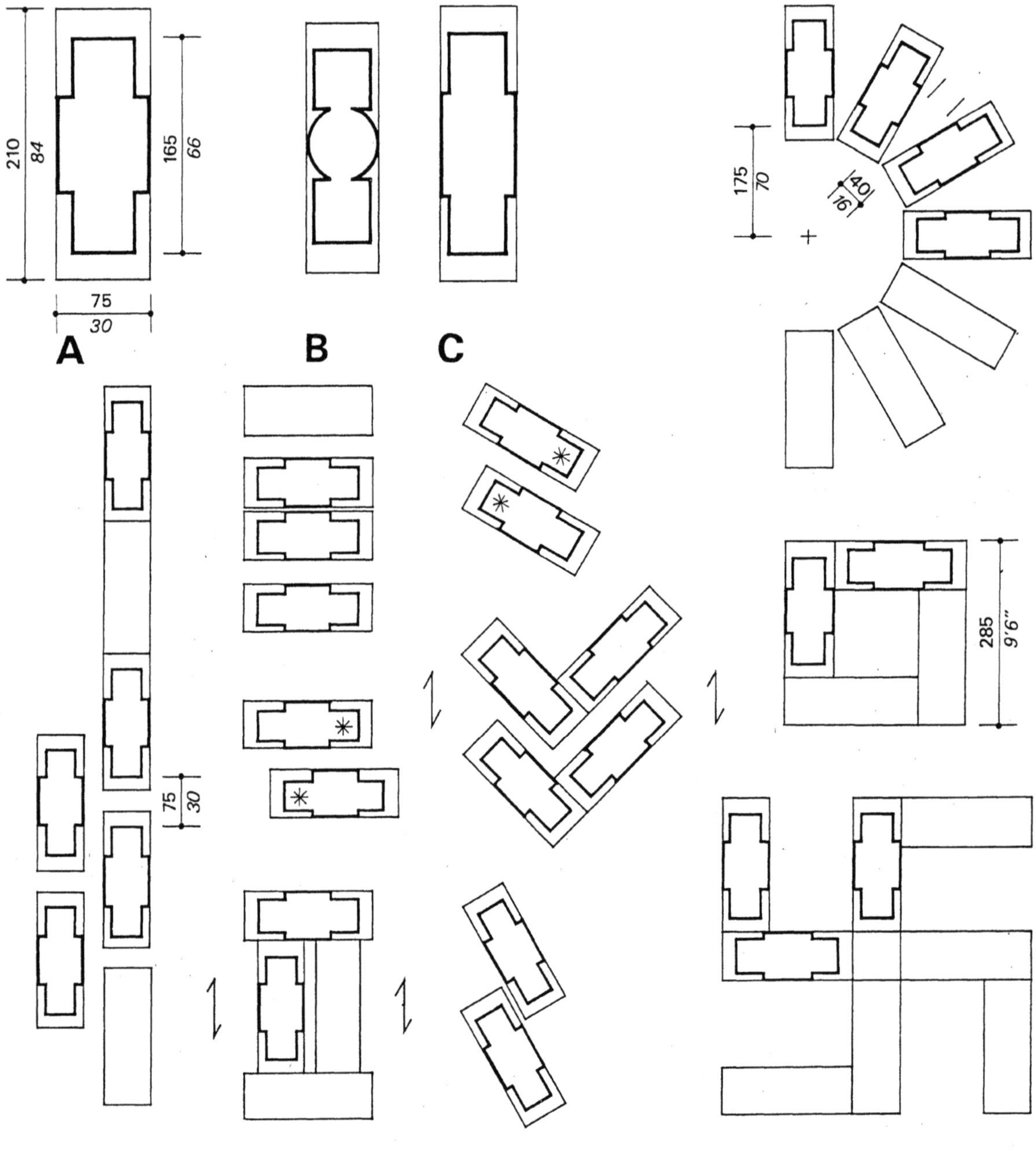

1:50,100

- When two or more individual seating assemblies (footprints and modules) are grouped together, this larger configuration – whether composed of identical or different units – may be termed a **Formation**.
- In combination, the assemblies may be repeated, alternated, mirrored, interlocked and set out radially. And the resultant formation may be categorized, *inter-alia*, as:
  line astern (aligned longitudinally);
  line abreast (aligned laterally);
  echelon (stepped)]
  herring-bone (alternate rows at contrasting angles);
  windmill (particular form of "rotating" figure).
- These written descriptions effectively cover the various formations that appear on this and the next five pages. They are not idly or pedantically paraded, for words and diagrams share a common purpose: *to trigger the designer's imagination.*
- The formations illustrated examine "like with like", are mostly self-explanatory and call for few comments. Some, it will be appreciated, are less practical than others.

A Basic unit: 75cm/*30"* square table.
B, C Implied alternatives: 60cm/*24"* diameter and 60×80cm/*24"×32"* rectangular tables.

- Seats shown asterisked: a cozy tête-à-tête could be endangered by such close eye-to-eye distraction.

5.14

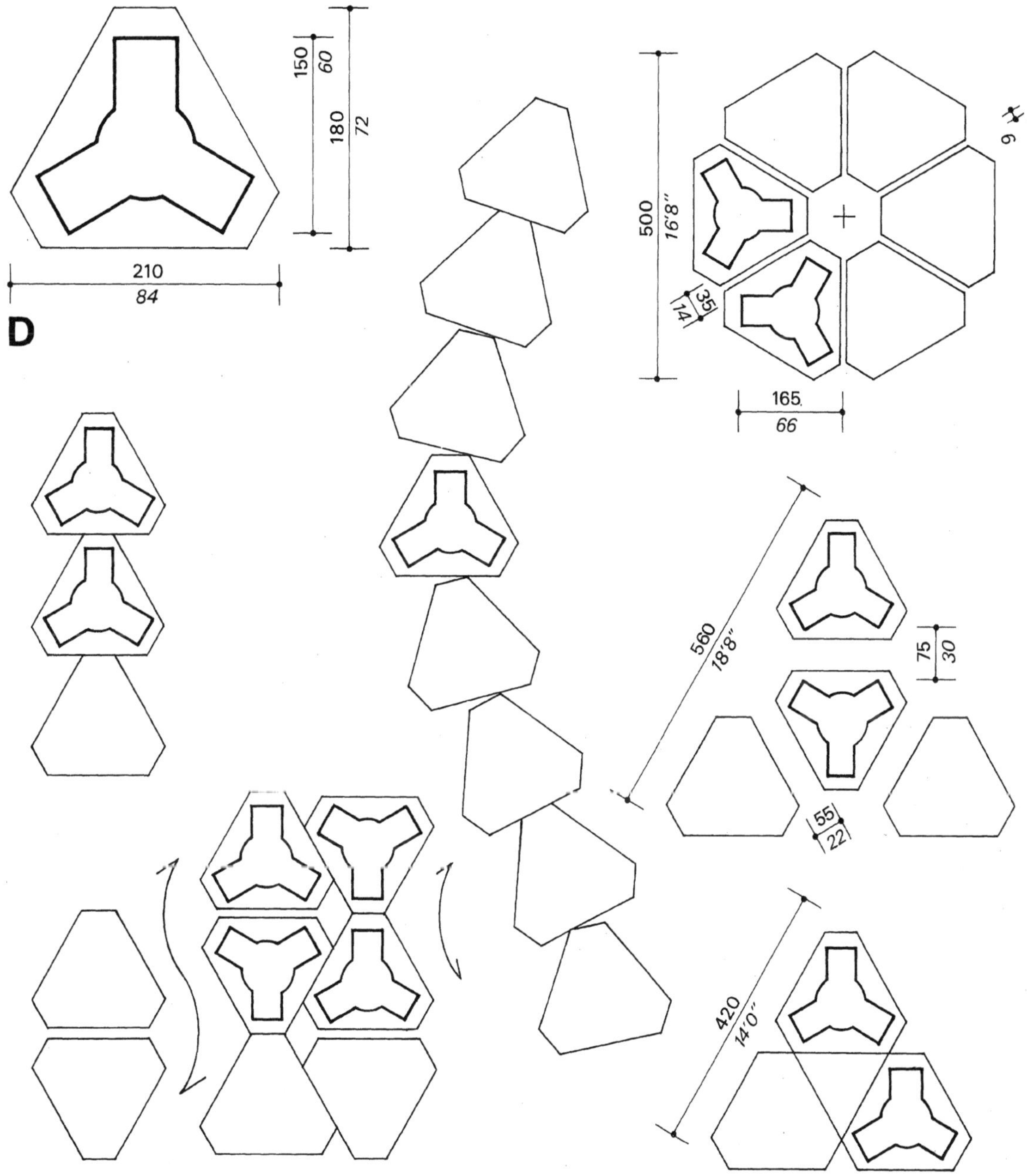

1: 50,100

An outline shape that encourages interlocking and change of aisle direction.

D Basic unit: 80cm/*32"* diameter table.

MODULES AND PLANS
Formations: rectangular tables, seating 4

5.15

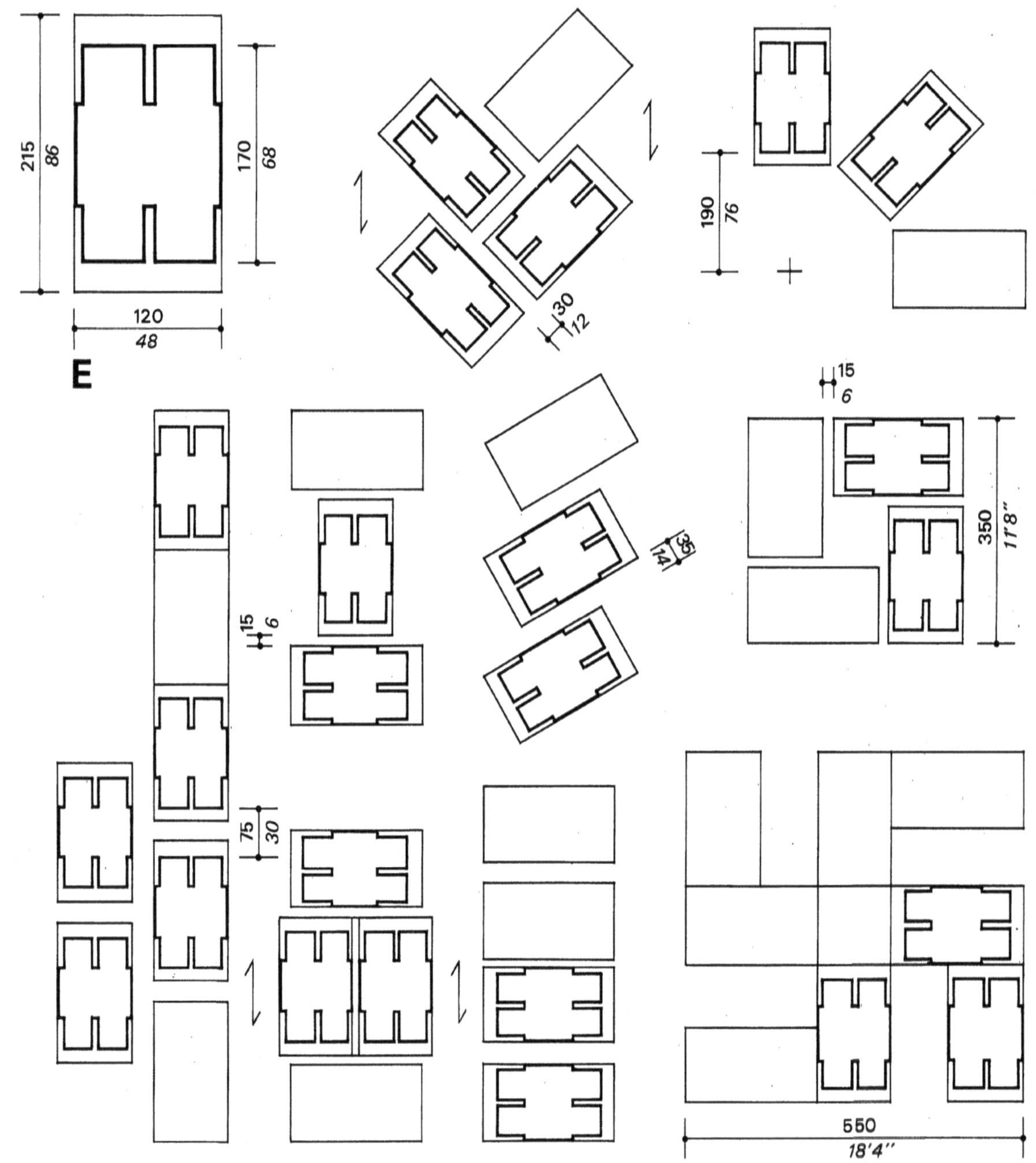

1: 50,100

- A most versatile footprint.

E Basic unit: 120×80cm/*48"×32"* rectangular table.

MODULES AND PLANS
Formations: round and square tables, seating 4

5.16

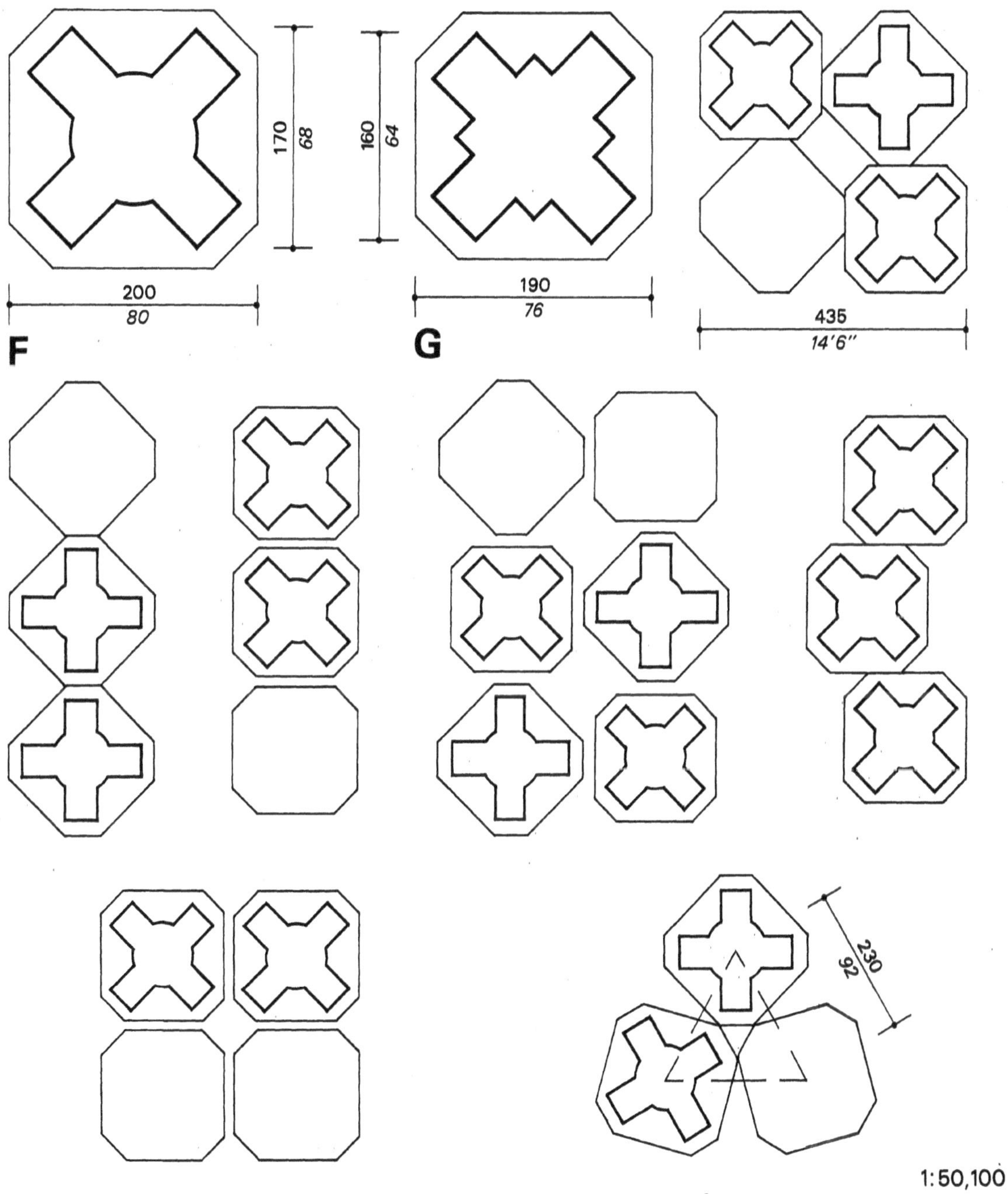

- The most common, interchangeable, 4-seaters. Spatially most efficient when set out diagonally in relation to the aisles.

F, G Basic units: 100cm/*40"* diameter and 90cm/*36"* square tables.

MODULES AND PLANS
Formations: triangular-based tables, seating 2, 3 and 4

5.17

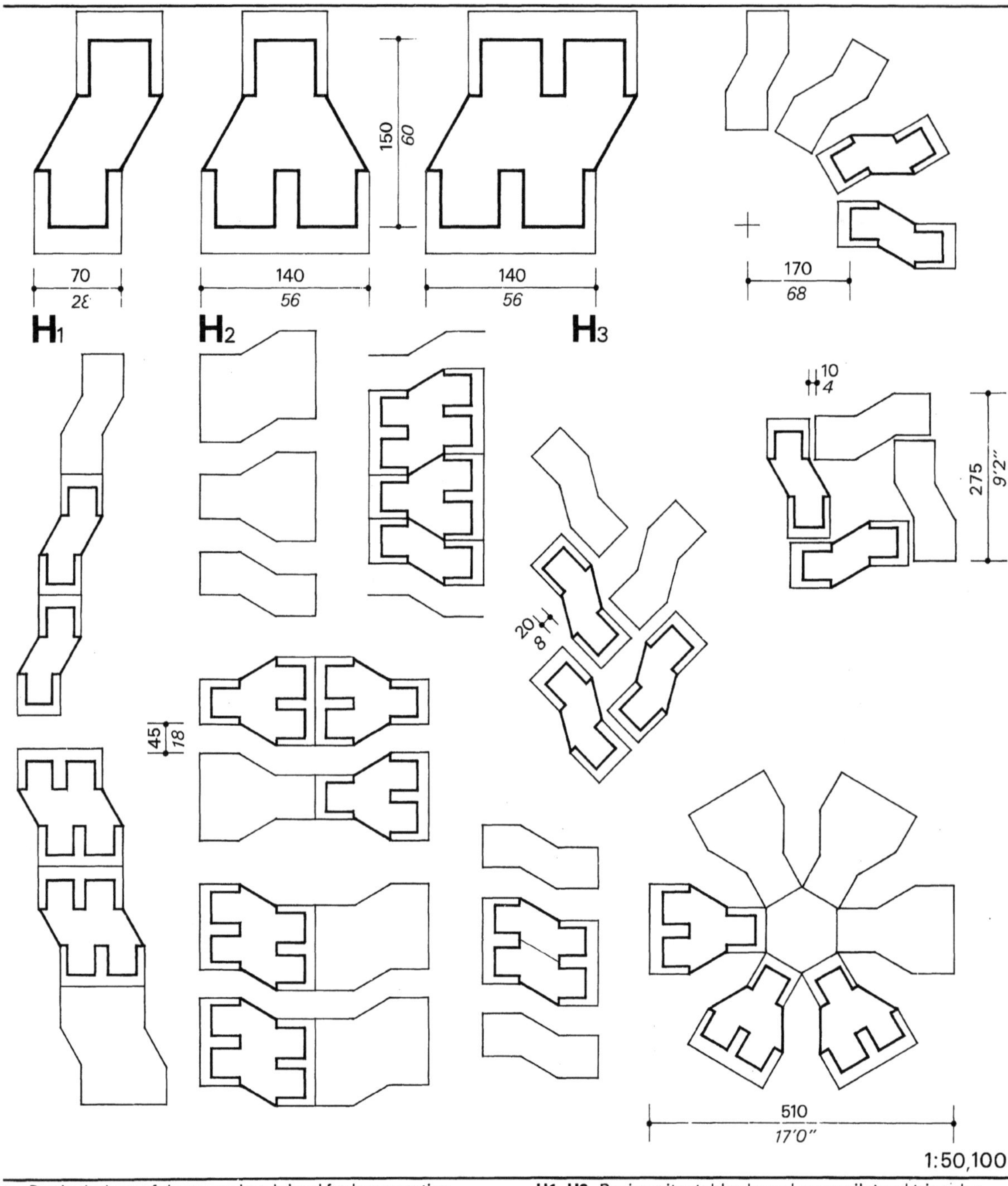

- Particularly useful range when joined for large parties of diners, or for creating angled aisles.

H1–H3 Basic units: tables based on equilateral triangles of 70cm/*28"* sides.

Formations: table/seat units, seating 2–6

5.18

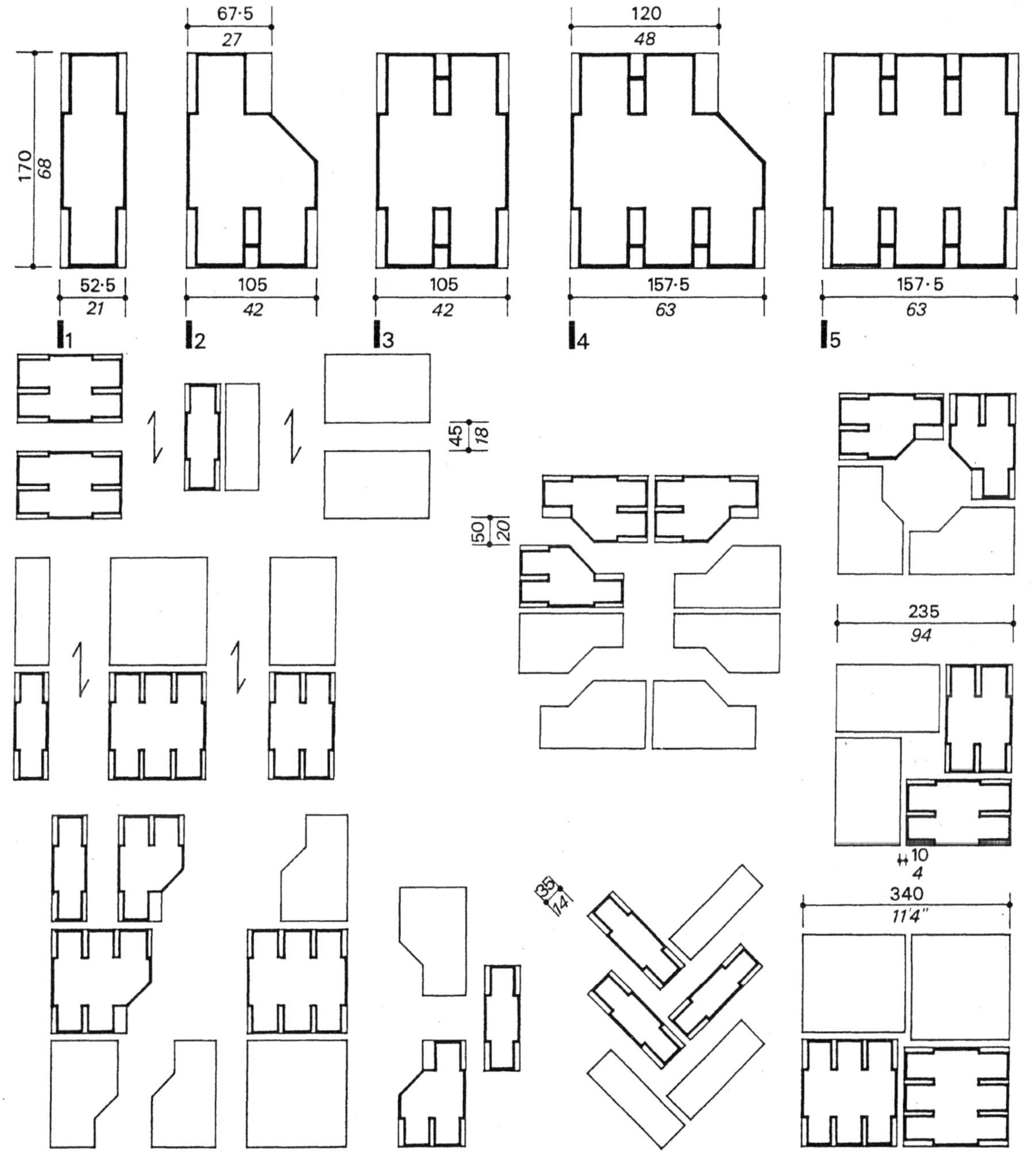

1:50,100

- Varying the table sizes over a small area (see bottom left) can produce angled aisles.
  Four- and six-seaters combine exceptionally well in "windmill"formation (see bottom right), normally with intervening screens.

I1–I5 Basic table/seat units: dimensions reflect a constant 75cm/*30"*table width, and table lengths in multiples of 52.5cm/*21"*per "place."

Plans: width factors

5.19

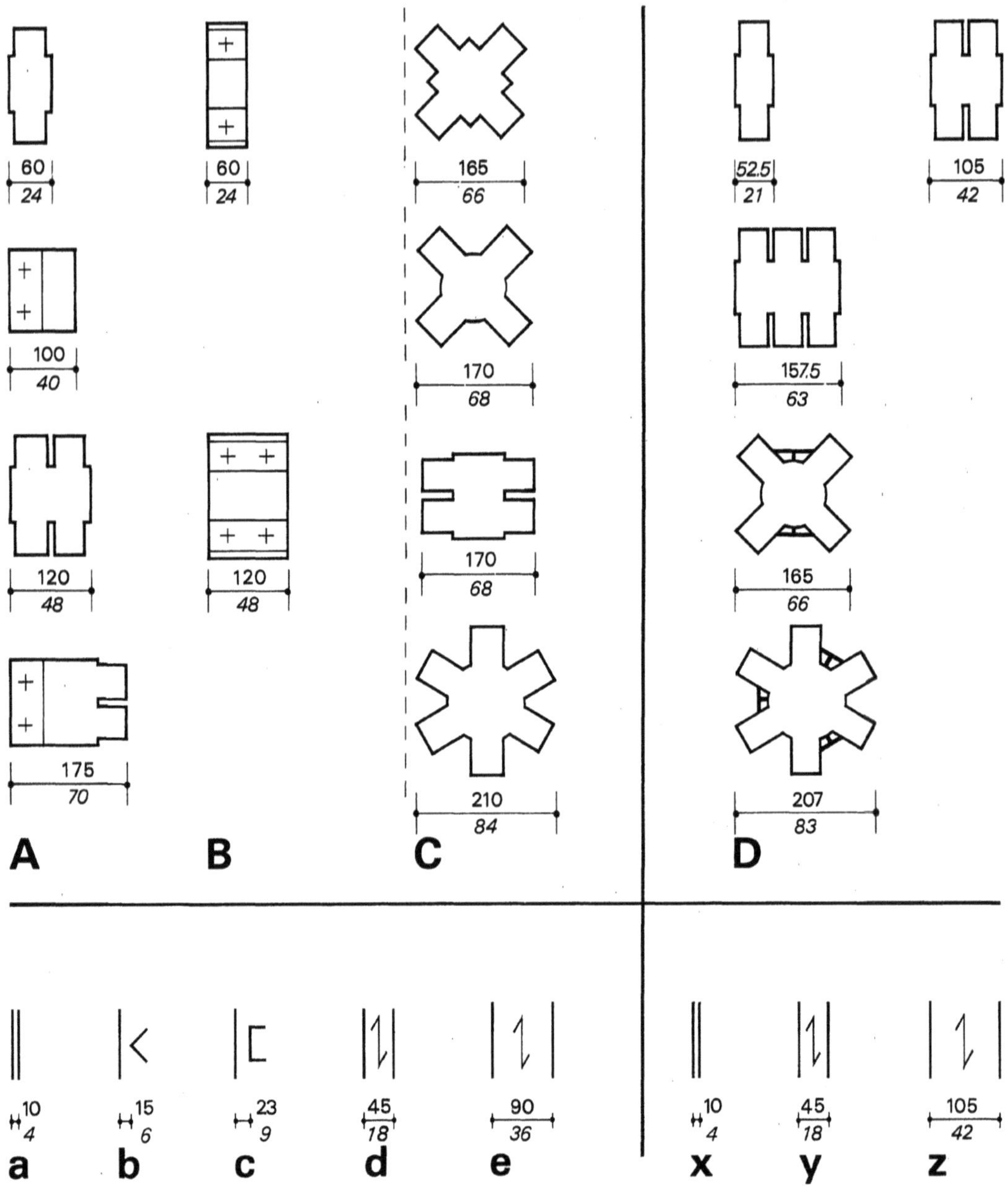

1:100

- The average restaurant is rectangular, and deeper than it is wide. Since the main aisle will always run from front to back, it is the lesser *width* dimension that must primarily govern the initial planning.
- This may not seem so important in very large eating establishments, or those located on a street corner or free of other buildings. Economic pressures, however, are likely to favor a predominantly rectilinear layout, and striking the right seating balance still suggests analysing width factors in the first instance.
- As an aid to this analysis, we show a "starter kit" of the most common seating assemblies for 2, 4 and 6 together with some clearance widths. (Conventional furniture is shown on the left, combination units on the right.)

- All dimensions are as previously shown, and all *footprints* are indicated at right angles to imaginary aisles, or walls, running lengthwise up the page.

A Suitable tables for wall location.
B Equivalent booths.
C Free-standing tables.
D Table/seat units.
a, x Minimum gap for divider screen between adjoining tables.
b, c Seat access clearances.
d, y Local customer route.
e, z Main aisle.

5.20

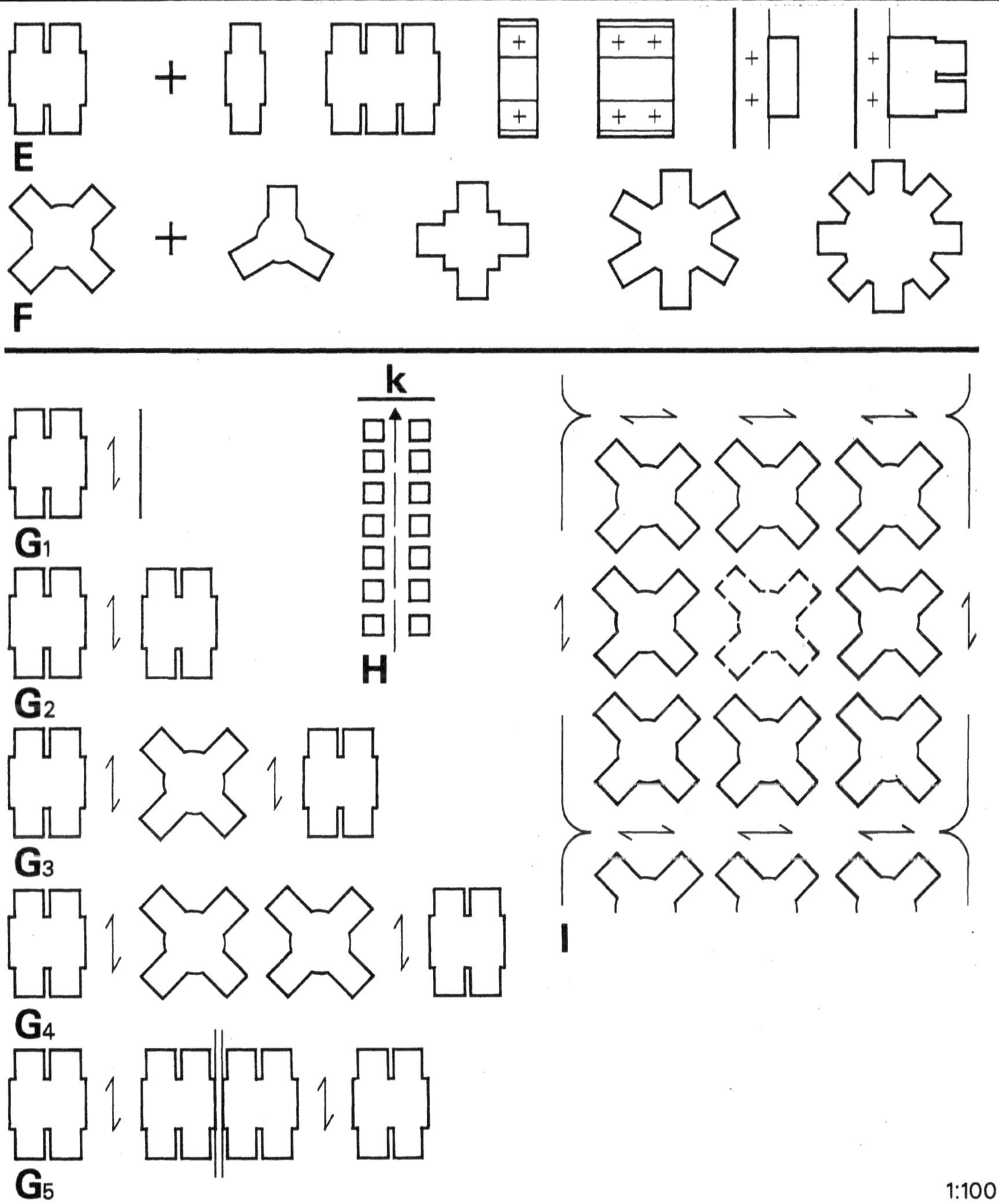

- There are two distinct categories of seating arrangement: those particularly suited to edge locations (E) and those that are not (F).
- Equally clearly, a restaurateur's major objective – quality of food and ambience, aside – will be to provide as many covers as space will allow. In a typical deep plan entered from one end, a maximized ratio of seats to circulation can effectively be achieved by: treating the available width as a series of *transverse bands*: and eliminating all possible waste space.
- Waste space will obviously be minimized if:

a footprints are compact in relation to their context.
b between-aisle seating consists of two rows rather than one. Three rows are impractical: see (I).

- Basic layout alternatives may be portrayed with two types of 4 person table – abutting (E) and free-standing (F). Progressively illustrated (series G) or any notional width band, the alternatives may be expressed as:
E–, E–E, E–F–E, E–F F–E, and E–E E–E.
- In terms of efficiency – number of seats divided by overall width – G2 is the best, G1 the worst.

E Rectangular table for 4, plus other *abutting* units.
F Round table for 4, plus other *free-standing* units.
G1–G5 Basic layout alternatives, in width bands.
H Typical deep plan (diagrammatic).
I 3-row plan, showing *central inaccessibility*.

Plans: typical room widths

5.21

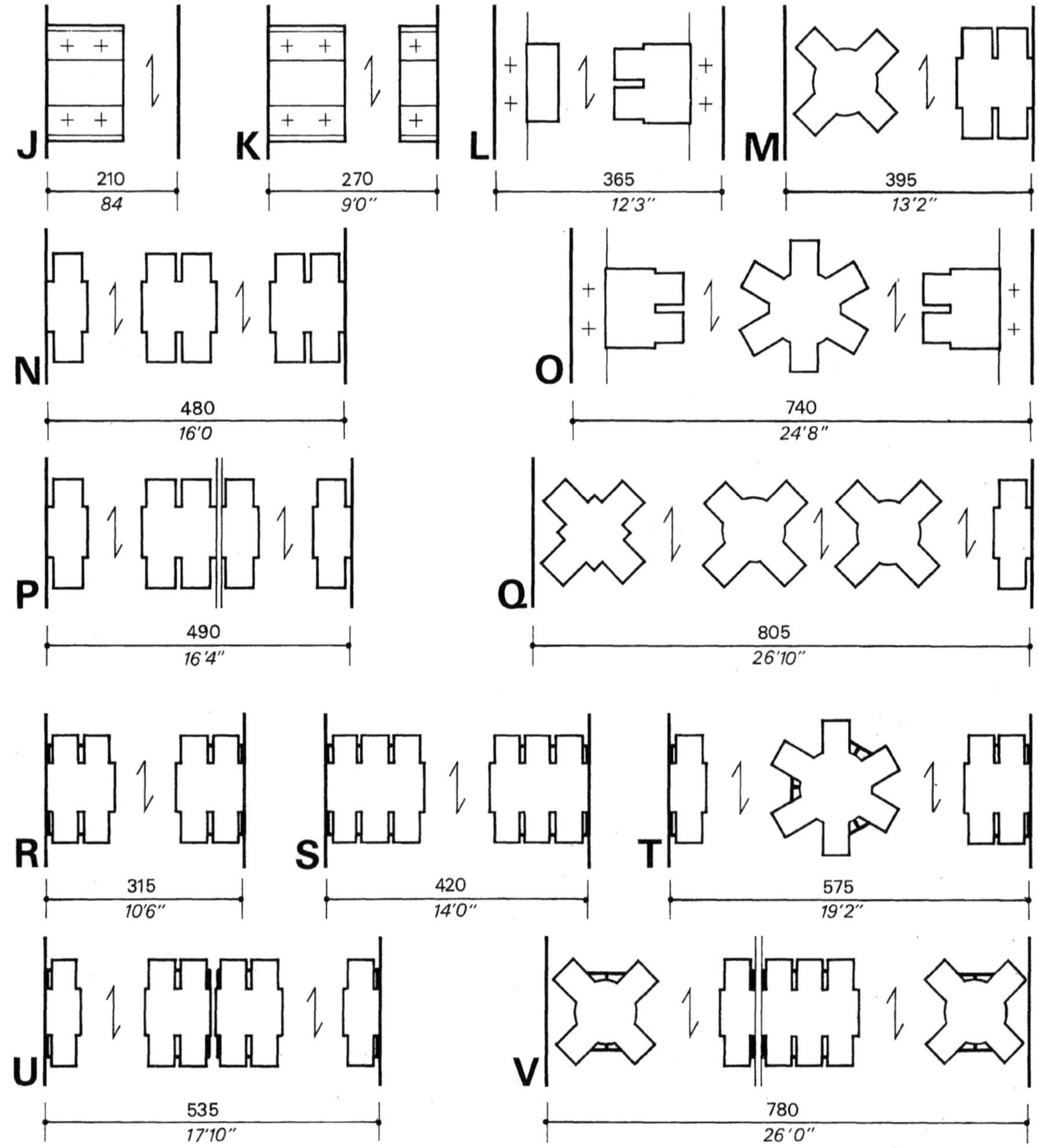

1:100

- This sheet continues the exploration of layout bands, introduced in 5.20, and exhibits a wider range of seating and overall width.
- The top three rows of diagrams (**J–Q**) show conventional furniture, the lower two (**R–V**) table/seat units. The different settings range from a minimal single-sided aisle configuration (**J**), as often found in narrow sandwich bars, to double-band twin aisle layouts (Q, V) occupying a width of around 8m/*26'*.
- The importance of *mix* – tables seating different numbers of people – should not be overlooked, and most examples demonstrate that variety may neatly be accommodated within the same band.
- Two typical workings will suffice to show how these overall dimensions are made up, and how the local density may be calculated:

Q 15 (margin) + 165 + 90 (aisle) + 170 + 45 (access) + 170 + 90 (aisle) + 60 = 805.
Depth = 200 (module), no. of seats = 14. Area = 805 × 200 = 16.1m^2.

Density $= \frac{\mathbf{16.1}}{\mathbf{14}} =$ **1.15m^2/*12.4ft^2* per person.**

V 10 (margin) + 165 + 105 (aisle) + 53 + 10 (screen) + 157+105 (aisle) + 165 + 10 (margin) = 780
Depth = 200 (module + margin), no. of seats = 16.
Area = 780 × 200 = 15.6m^2.

Density $= \frac{\mathbf{15.6}}{\mathbf{16}} =$ **0.98m^2/*10.5ft^2* per person.**

5.22

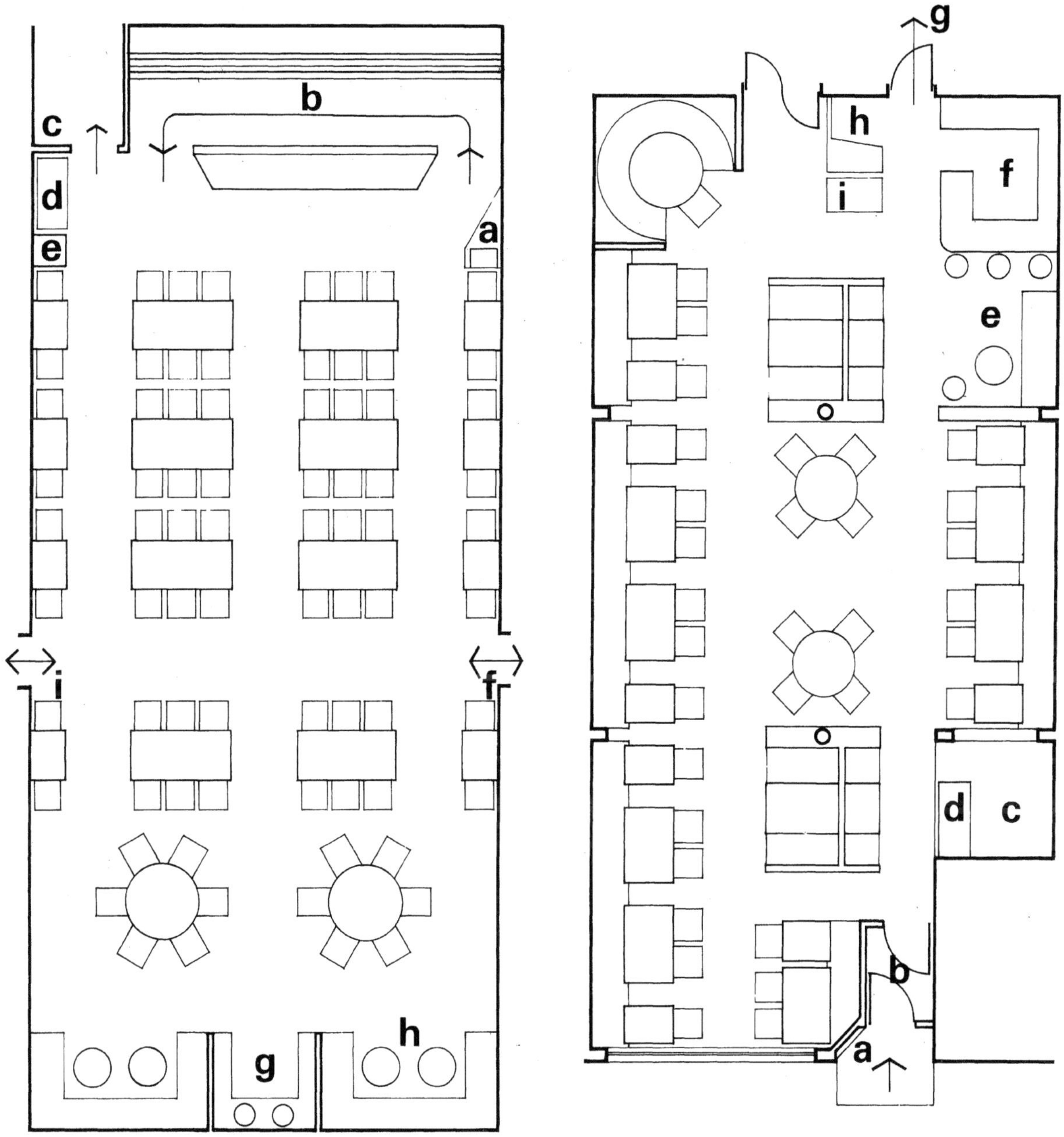

1:100

Hypothetical layouts, for a corporate cafeteria (*left*) and downtown restaurant (*right*), that exhibit some of the arrangements and guidelines displayed earlier. Some arbitrary assumptions, including a common overall dimension of 15m×7.5m/*50′×25′*, were made before turning to the interior planning.

CAFETERIA (*left*): clear-span top floor of a new office block, entered from one side and with a terrace on the other. Available width suits 2- and 6-seater tables rather than 4s. Siting two round tables and the coffee area at the rear dispels any sense of institutionalism.

Occupancy: 76 *dining* 112.5m^2/*1286ft^2* gross (*including* coffee zone), i.e. a density of 1.48m^2/*16.9ft^2* per person.

RESTAURANT (*right*): at street level, with upper floors carried on columns and piers and independently accessed. Inevitably a two-aisled scheme. Emphasizing the structural bays with low dividers and built-in booths helps to meet the need for variety.

Occupancy: 74 *dining* in 106m^2/*1210;ft^2*, i.e. a density of 1.43m^2/*16.4ft^2* per person.

Cafeteria: a tray pick-up, **b** counter, **c** dish return, **d** clearing trolley, **e** trash bin, **f** entrance, **g** tea/coffee, **h** coffee lounge area, **i** fire exit.

Restaurant: a menu display, **b** draft lobby, **c** coats, **d** dumbwaiter, **e** reception, **f** bar, **g** toilets, fire exit, **h** head waiter's station, **i** sweet trolley.

MODULES AND PLANS: summary

1 An individual assembly of table and chairs may, for convenience in considering its outline, be termed a ***footprint***. Adding the small spaces needed to get in and out of the seats, gives the footprint its immediate territorial ***envelope***, or ***plan module***. If only for descriptive purposes, a yet further step may be noted; and this occurs when two or more modules are so closely grouped as to constitute a ***formation***.

2 Given an overall space to be filled and a specified mix of seating, the designer can then manipulate modules and formations – together with the necessary aisles – and produce an initial layout. Theoretically, and put at its simplest, this procedure summarizes the design process.

3 Many factors help to shape, and complicate, this process, the ways in which modules combine, their alignment and efficiency (comparative density), the placing of entrances and exits, the location of any isolated supports, and the available overall dimensions.

4 Access, whether that of a customer to his seat or a server to the tableware, is another major consideration. Suitable aisle widths and local clearances can be established to meet the anticipated traffic; and, as they are seldom bounded by continuous straight lines, aisles need not necessarily maintain a constant clear width.

5 Insofar as they open up new avenues, the conscious grouping of modules into formation patterns can be a rewarding exercise. Formations such as "line astern," "line abreast," "echelon," "herring-bone" and "windmill" are sufficiently widely used in other contexts, e.g. the armed forces, to be self-explanatory.

6 Finally, selecting the appropriate seating arrangements and forcing them into actual plans will be assisted by recognizing that:

- Seating modules fall into two categories: those particularly suited to edge conditions, and those that demand to be free-standing.
- A patron or restaurateur will, in general, expect the highest ratio of seats to circulation and waste space that is consistent with good planning.

The next, concluding Section takes a brief look at some situations that fall outside the commercial mainstream.

SPECIAL SITUATIONS

6

6.01

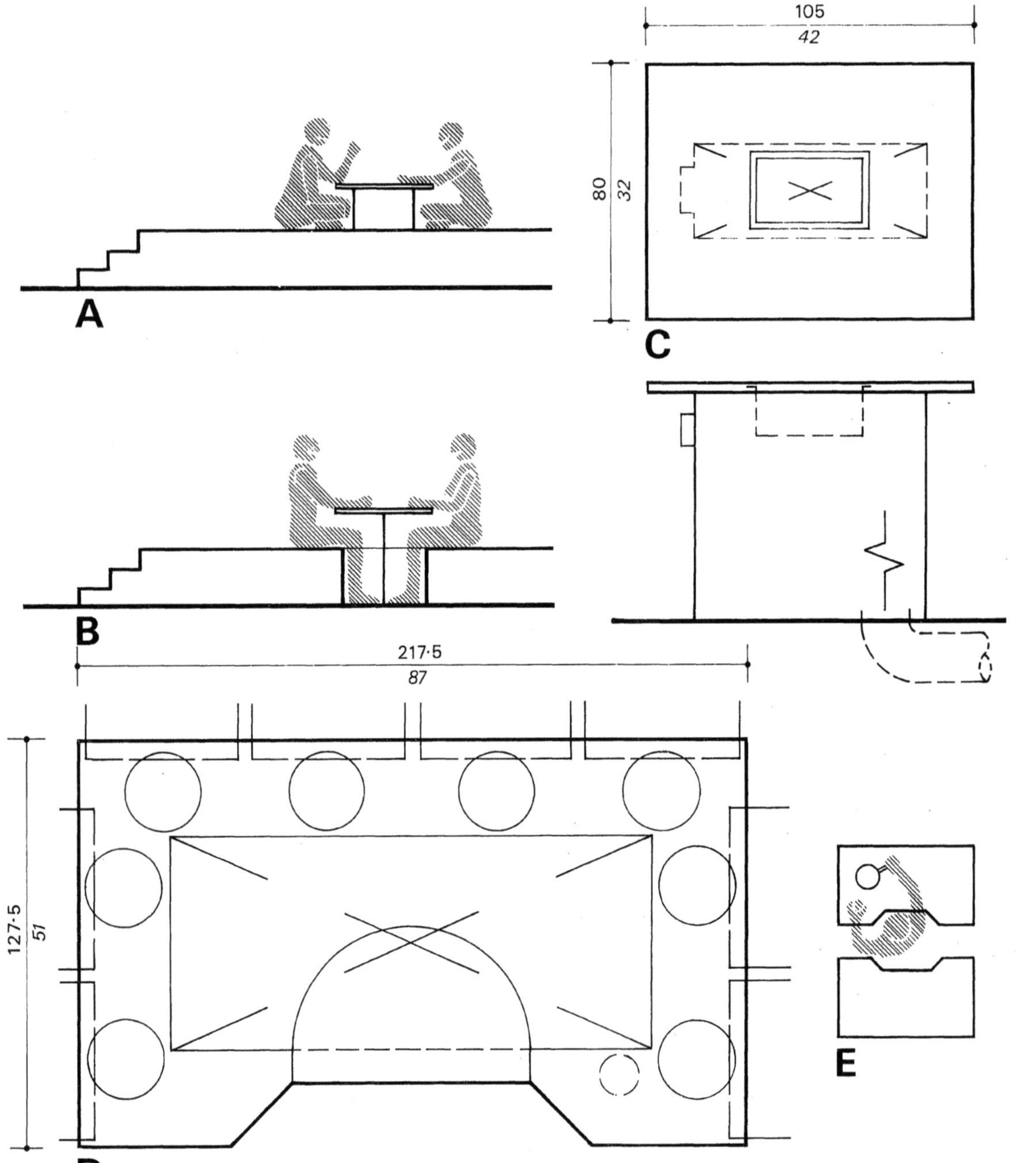

1:20,50

- For many Westerners, to dine in oriental style is a welcome change. The Korean and Japanese examples, shown here, call upon long-standing tradition. Though different, they share a sense of pre-ordained formality that distinguishes such settings from the average, universal restaurant.
- Removing shoes, ascending a raised platform, sitting cross-legged or around a foot well – these are but the preliminary observancies of a ritual that celebrates cooking at the table. Such ritualistic experience may be epitomized by dining at a Japanese *tappan-yaki*: an exact complement of eight "guests", seated around a hotplate and served by their own chef.

A Traditional Korean setting, table on a platform.
B Later version, feet resting on main floor level.
C Contemporary 4-person Korean "cook at" table incorporating a sunken grill with gas-fired burner and mechanical extraction to underfloor ducting. Note restricted footroom.
D Japanese *tappan-yaki*: table top inset with a stainless steel hotplate, part of which is heated to cooking temperature, above a storage cabinet. Note the close seating.
These tables are invariably positioned in back-to-back pairs (**E**), with the chef turning his attention to the other table for the second sitting.

SPECIAL SITUATIONS
Dining formally: refectory banquet layouts

6.02

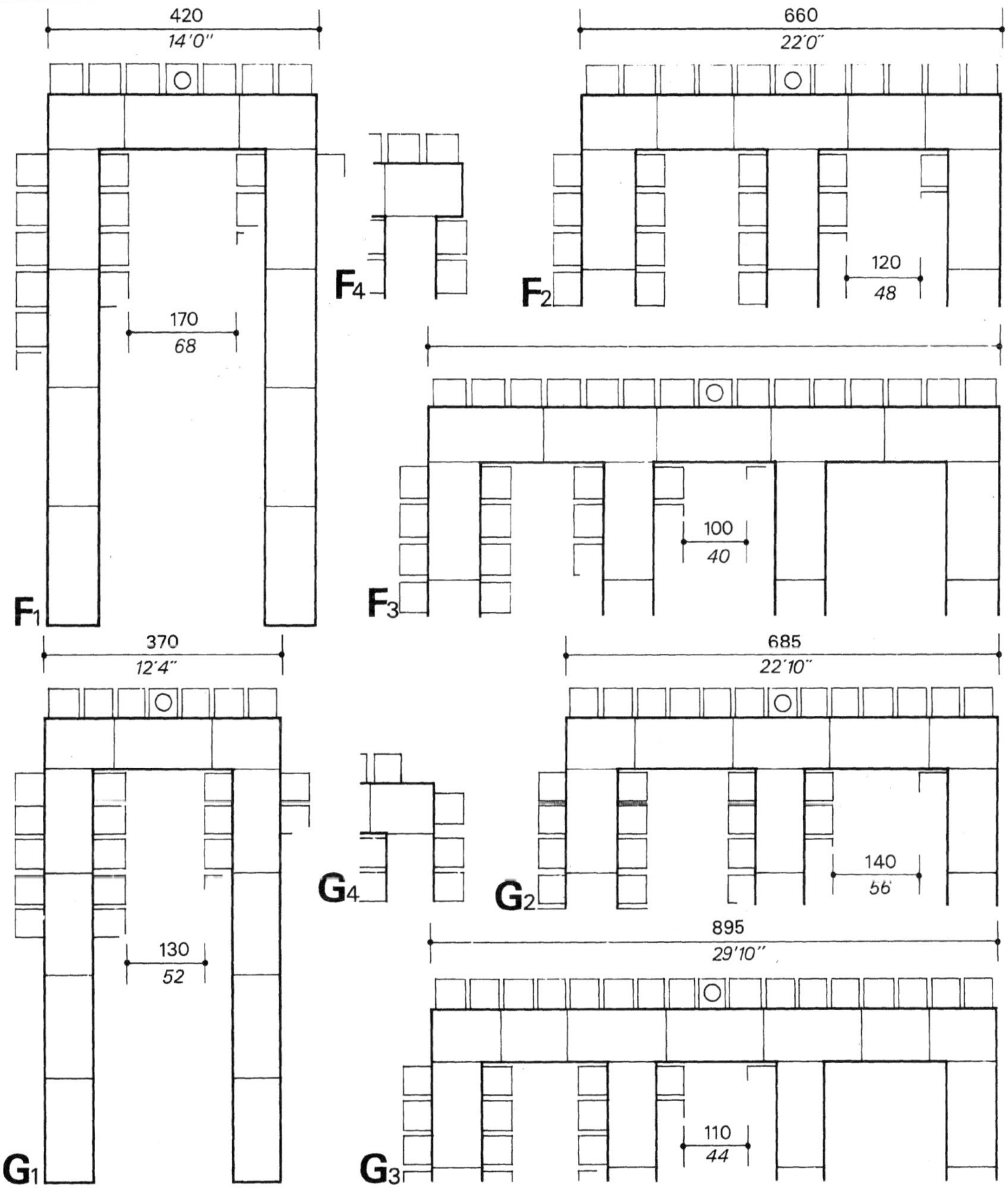

1:100

- Banquets are traditionally laid out with a long centrally-presided head table and one or more "offshoots" at right angles to it. Depending on circumstances and available space, the overall refectory configuration is likely to be T-, U- or E-shaped.
- The various alternative layouts (and part-layouts) illustrated are based on a *bi-modular* set of short tables, i.e. of two different lengths, as such a system will precisely fulfill any required number of seats in a given run.

With so many people dining together, rapid and efficient waiter service is paramount. Potential congestion lies within the fingers of the plan, and the distance between parallel table edges should not fall below 180cm/*72"*.

F1–F3 Layouts based on 80×120 or 180cm/32"×48" or 72" tables.

F4 Alternative corner-turning.

G1–G3 Layouts based on 75×105 or 157.5cm/*30"×48" or 63"* tables.

G4 Alternative corner-turning.

Dining formally: round-table banquet layout

6.03

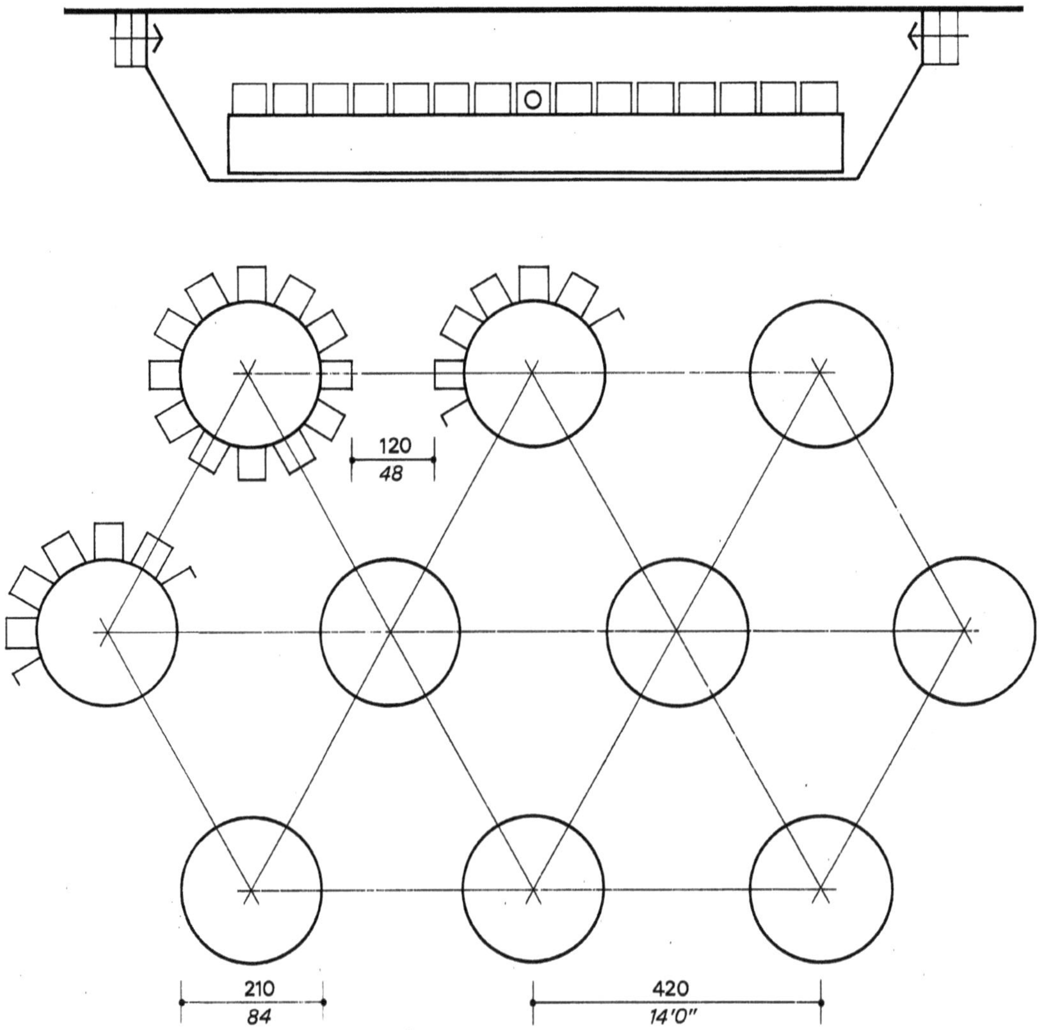

1:100

- Another banqueting model is that of a head table, often raised on a dais, and several big round tables seating ten or more persons each.
- Shown, in this example, are 12-seat tables set out on a hexagonal grid. The resulting network of generous aisles will be capable of handling large numbers of guests and staff.
- Such tables are also often used in conference centers and congress halls.

Dining formally: domestic dinner party, seating 8

6.04

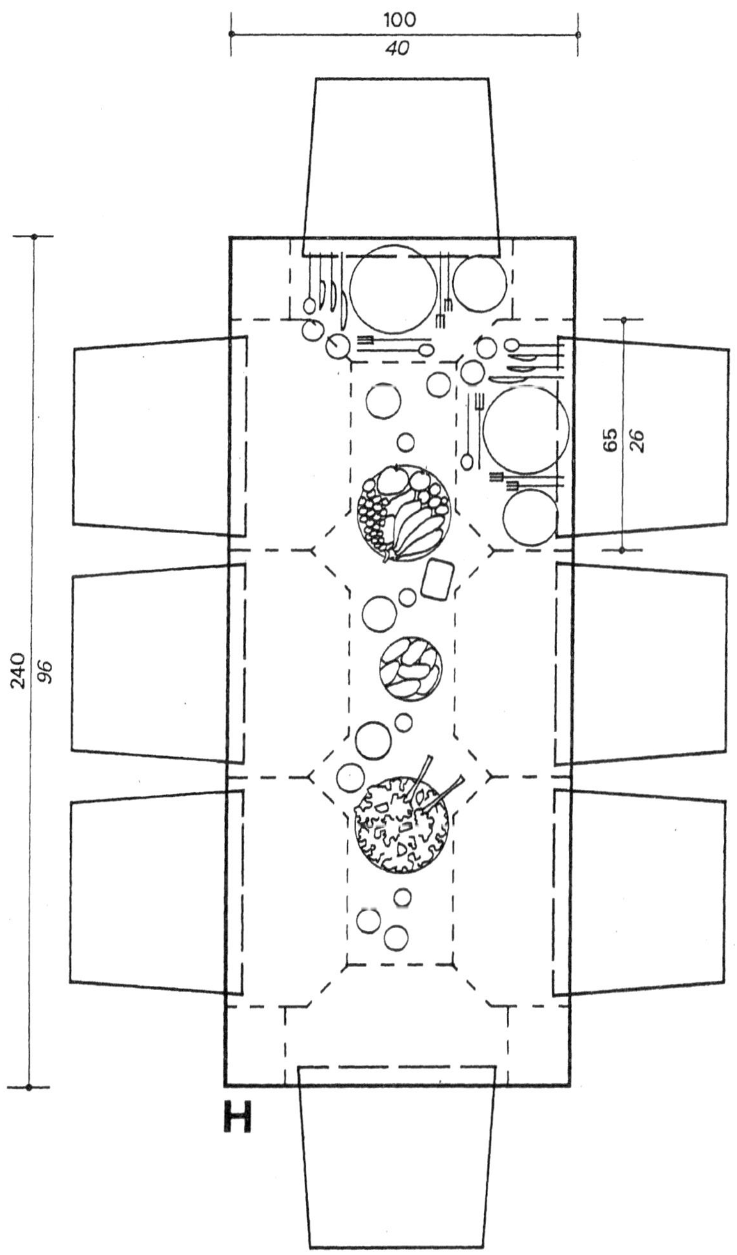

H

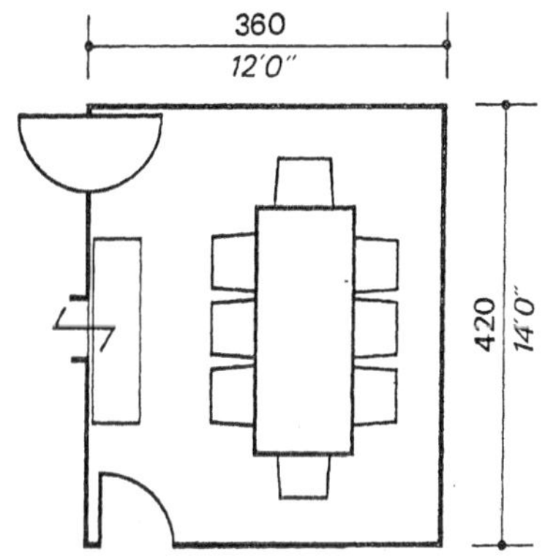

I

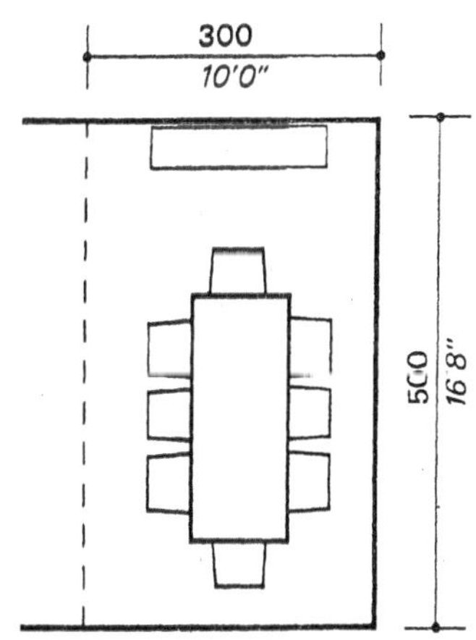

J

1:20,100

- Lavish, relatively formal, hospitality – or, say, a family Christmas dinner – demands a higher degree of comfort and space than is provided by a normal restaurant. Those in a position to do so will naturally equip themselves with the most generous table that their dining room can take.
- Indeed, in this regard comfort *is* space: space for spreading elbows and legs, and for all those communal dishes and tableware items. Such a table for eight will require the sort of dimensions shown in H. Allowing for a side-board, on which further dishes can stand or the joint be carved, sets a minimum dining room/dining area of 15m²/*160ft²*.
- For larger rectangular tables, seating upwards of 10, allow 65cm/*26"* extra length per 2 additional persons and an appropriately increased dining room.

H 240×100cm/*96"×40"* table, with 65cm/*26"* place settings and armed chairs approaching 60cm/*24"* in width. The central surface is given over to candlesticks, salad bowl, bread-basket, water carafe, wine bottles, fruit, butter dish, and cruets.

I Suitable dining room, with door to kitchen, 50cm/*20"* wide sideboard, and serving hatch. Size determined by furniture, 75cm/*30"* clearance between chairs and sideboard, and 45cm/*18"* clearances elsewhere.

J Suitable integral dining area, with identical furniture.

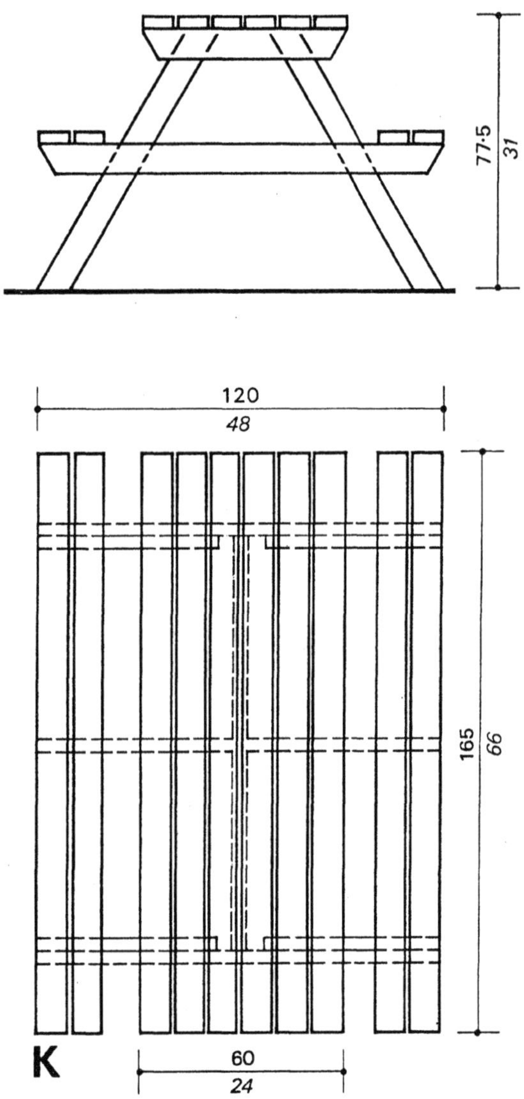

K

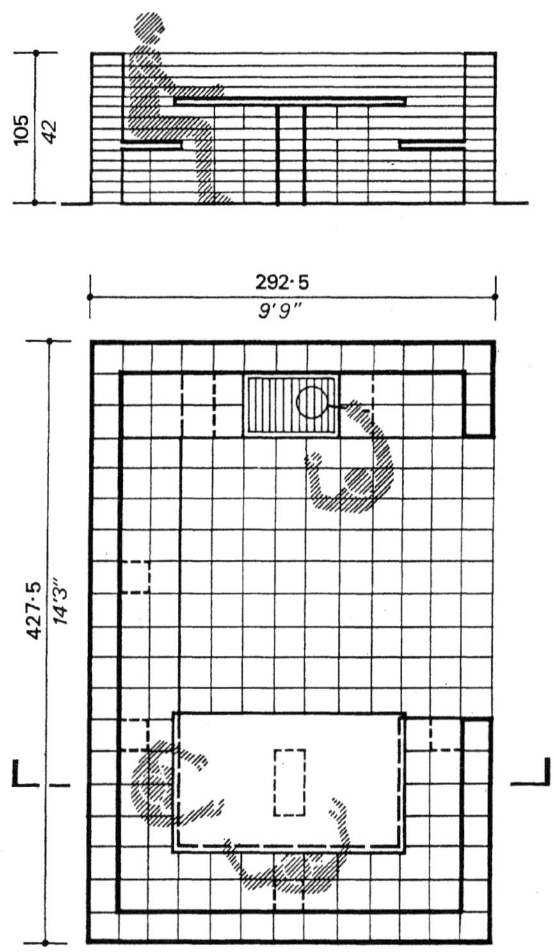

L

1:20,50

- *Dining* outdoors, with conventional table and chairs, makes the same anthropometric and functional demands as dining indoors. *Eating* in the open requires no such standards because this activity may well be incidental to the scenery and weather. A packed lunch can be enjoyed standing up, a spread picnic knows no bounds.
- However, midway between the conventional and unconventional extremes are certain specific solutions to the question of eating and drinking outdoors.
- Two of these are shown: an A-frame bench-table – to be seen in picnic sites and private gardens and on pub terraces – and a low-walled enclosure designed for barbecues but equally suited to other social gatherings. Both are unaffected by the elements.

K Bench-table: sturdily engineered from the same size of timber. Gaps between the planks allow rainwater to disperse.

L Brick enclosing wall, built-in table and wrap-around bench, with a barbecue grill, associated worksurfaces and fuel storage below. Dimensions are based on standard bricks but alternative materials would include stone and precast concrete, and slate tops.

SPECIAL SITUATIONS
Eating on the move: train / camper

6.06

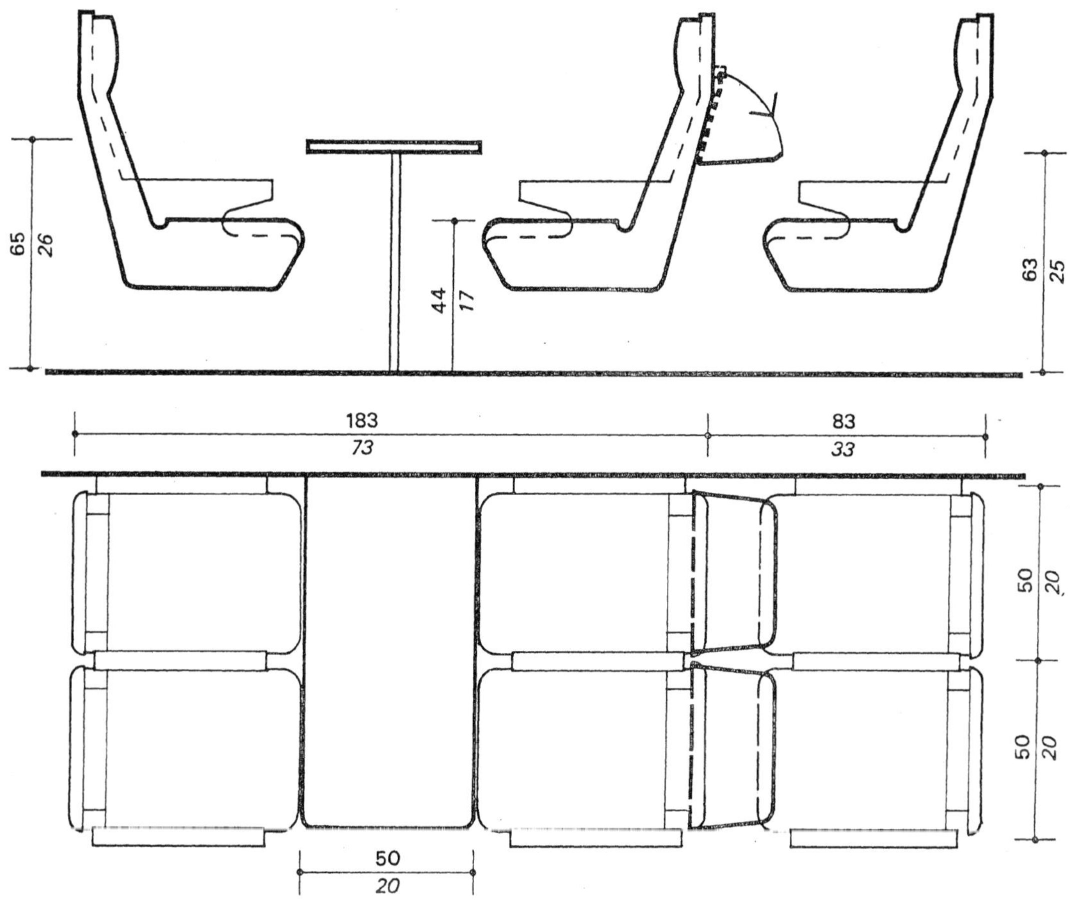

M

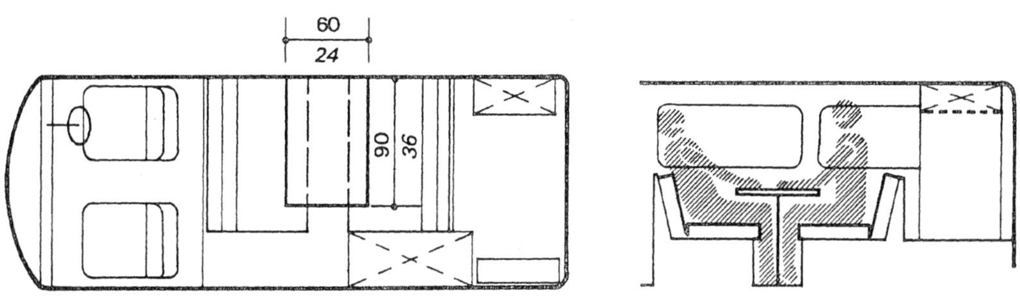

N

1:20,50

- Feeding in a ship's saloon, say, or railroad dining car is analogous to dining conventionally while stationary.
- But, whether traveling on wheels or in the air, eating at one's *passenger seat* is an intermittent and secondary activity for which the already tight space must, at best, be adapted. Two examples (**M** and **N**) will serve to illustrate the necessary functional flexibility:
- Seat backs, in a plane, are never wasted. In addition to pockets for in-flight information, brochures, emergency instructions, sick bags and other necessities, there is a hinged flap for drinks and meal trays. Customarily little larger than 45×25cm/*18″×10″*, it may be brought into play at any time during the journey. When folded back it takes up no room.
- Many train/railroad seats are now similarly fitted, as shown in M. Where seats are opposed in pairs there is enough space to insert a more normal table: frequently narrowing in width towards the gangway, for easier access, and always removable.
- Removable tables are also deployed in trailers and camper trucks.

M Hinged flaps, 48×26m/*19″×10.5″*, as fitted in a British Rail Western Region "Intercity" 2nd Class carriage.

N Seating arrangement in a VW Camper. The removable drop-leg table, normally stored below the high-level rear locker, may also complete the base to form a bed.

Eating casually: TV dinner / stand-up snacks

6.07

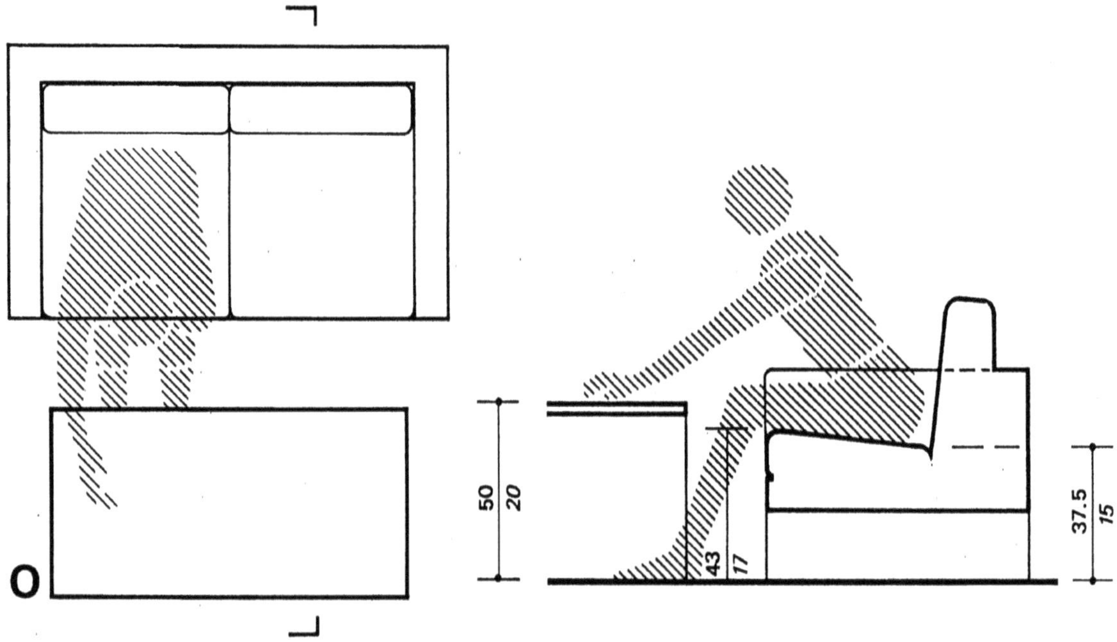

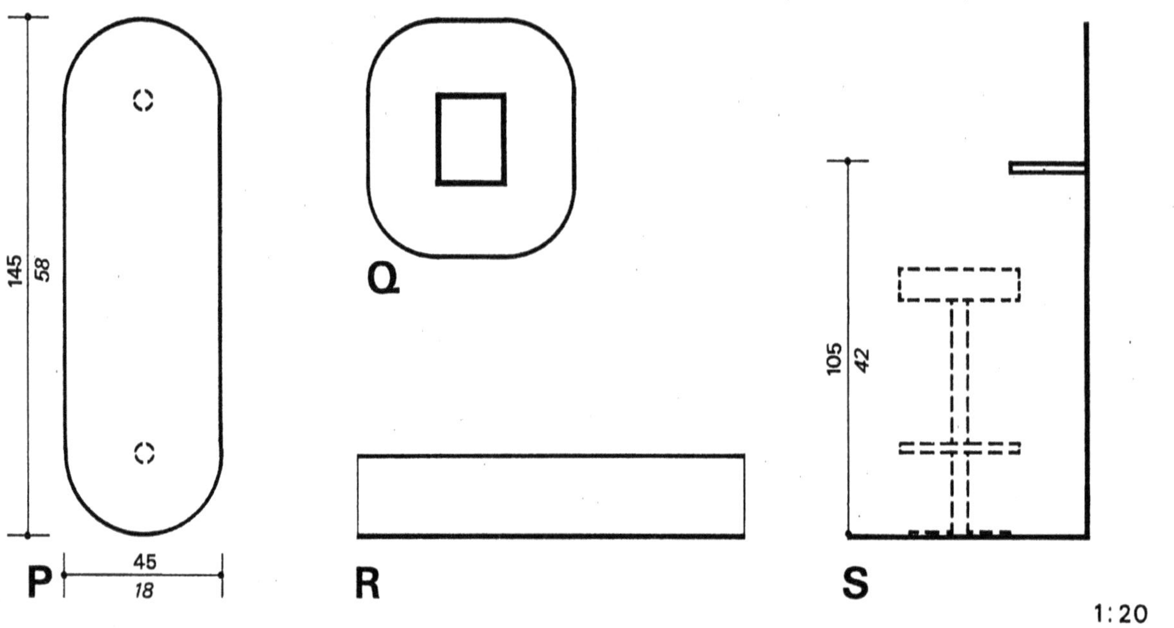

1:20

- Eating is relegated to the ultimate secondary activity in the situation where someone snatches a bite while comfortably seated on a couch, watching television.
- The dimensions and form of the furniture are established for relaxed sitting, and the height of the coffee table is in accord with this posture.
- When eating or drinking in our living rooms, or for that matter in bar lounges, we are prepared to adapt to the physical circumstances (**O**).
- Where space is scarce, notably in public concourses, small cafes and crowded bars, we adapt again: foregoing comfort for the sake of a hurried snack.
- There are three typical stand-up solutions: a high counter for up to six persons (**P**), a wall shelf (**R**), and one that encircles a free-standing stanchion (**Q**). The depth of such shelves should be at least 22.5cm/*9"*, and if the height of all surfaces is in the region of 105cm/*42"* this will permit any available high stool to be drawn up alongside.

SPECIAL SITUATIONS: summary

1 These random examples share a common distinction: to a greater or lesser degree, the meal itself is subservient to the event, location or both.

2 Formal dining introduces a transcending element – a sense of occasion, ritual, theatricality even – while eating informally is for the most part entirely secondary to the circumstances. Normal space planning considerations scarcely apply and, when they do, often run against expectations.

3 For example, one might suppose that banqueting or dining at exclusive oriental restaurants provides generous place settings, but the reverse is usually true. Given the need for maximum numbers, close-packed seating is what matters.

4 Under real pressure of space, such as in trains and airplanes or in a crowded bar, eating surfaces can be almost non-existent.

5 It is perhaps only in the private house or garden that people can spread themselves.

EPILOGUE

Following an exhaustive review of the details of dining and eating components and the spaces they occupy, and with some guidance on general strategy, certain observations are offered in conclusion:

- Very few dimensions, recommended here or elsewhere, are more than guidelines. Spatial relationships take precedence over exact measurement.
- The skilled architect or interior designer will know just when – and to what extent – he or she may bend the rules, to fit an all-embracing concept.
- It is in the province of the designer to add all the environmental and atmospheric touches (merely alluded to in this book) that may transform a bland workable solution into a memorable one. Individual flair is not prescribable.

FURNITURE DIMENSIONS
Comparative survey

Appendix 1A

ROUND TABLES Ø	2	3	4	6	8	10	12	persons
Published range	60–85 *24–34*	80–90 *32–36*	80–105 *32–42*	100–135 *40–54*	120–180 *48–72*	155–180 *62–72*	185 *74*	
As sized by formula			76 *30*	115 *46*	153 *61*	191 *76*	229 *92*	
Recom-mended	60 *24*	80 *32*	100 *40*	125 *50*	150 *60*	180 *72*	210 *84*	
SQUARE & RECTAN-GULAR TABLES l × w	2	4	2	4	6	8	6	8
Published range	68–78 *27–31*	85–105 *34–42*	68–90 60–75 *27–36* *24–30*	110–125 60–80 *44–50* *24–32*	160–180 75–80 *64–72* *30–32*	210–250 75–80 *84–100* *30–32*	145–180 75–90 *58–72* *30–36*	175–240 75–90 *70–96* *30–36*
Recom-mended		90 *36*	60 80 *24* *32*	120 80 *48* *32*	180 80 *72* *32*	240 80 *96* *32*	150 90 *60* *36*	210 90 *84* *36*
CHAIRS w × d spacing					**STOOLS w × dØ spacing**			
Published range	50–65 50–60 *20–26* *20–24*	45–65 45–55 *18–26* *18–22*	58–75 23–30		Published range	40–45 45–55 *16–18* *18–22*	30–40 *12–16*	42–70 *17–28*
Recom-mended			60 *24*		Recom-mended			70–80 *28–32*

RANGE – Taken from a broad selection of books, manufacturers' brochures and recommendations. As can be seen, this range is surprisingly wide and contains some inconsistencies. Several authors appear to have repeated sizes and spacings previously listed by others, without critical reappraisal.

FORMULA for sizing round tables. Often quoted, but an unreliable rule of thumb. Diameter is established by allowing 60cm/*24″* perimeter "cover" per person, and may be expressed (where p is the number of persons) as

$$D = \frac{60p}{3.14}$$

This formula approximates closely to reality for tables seating 8, but gives progressively inaccurate sizes for fewer or more people.

RECOMMENDED – Adequately generous dimensions, suited to the average restaurant, that recapitulate those proposed earlier in this work. With one exception (12-seat table), they fall within the published range. For obvious reasons, no recommendations are made for chair and stool sizes. The reader is however reminded that 50cm/*20″* square and 35cm/*14″* diameter, respectively, have been assumed for planning purposes.

CLEARANCES AND ACCESS WIDTHS

Comparative Survey

Appendix 1B

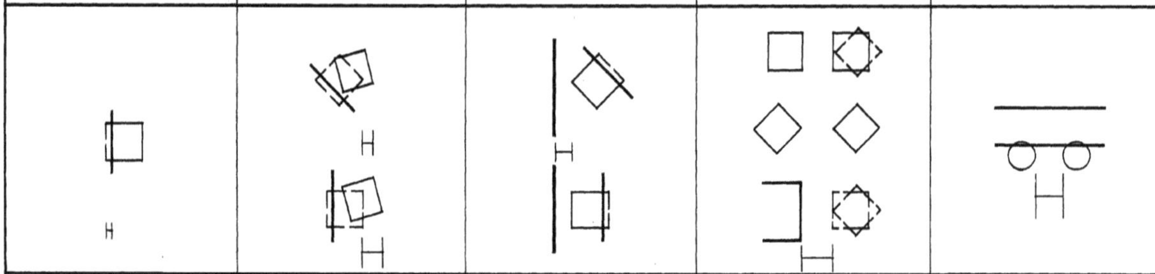

	table/seat overlap	sitting down & getting up	chair/wall clearance	chair/chair (or table) clearance	stool/stool clearance
A	3–10 *1–4*	10–30 *4–12*	15–30 *6–12*	40–60 *16–24*	12.5–25 *5–10*
B	5 *2*	15 22.5 *6 9*	22.5[1] *9*	45 *18*	35 *14*

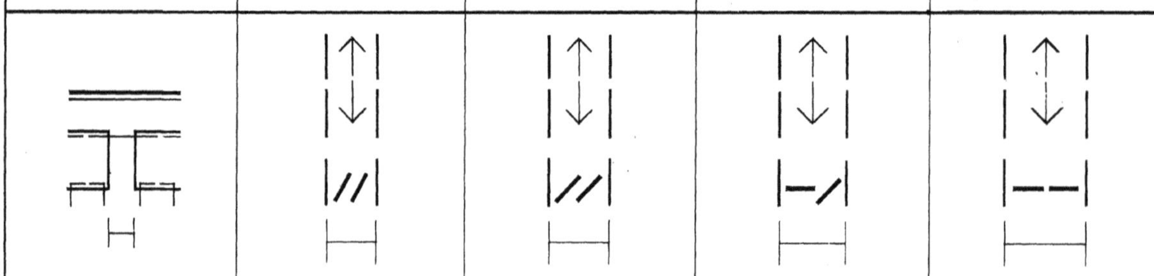

	bench seat access	LOCAL AISLE occasional service	LOCAL AISLE regular service	MAIN AISLE all-purpose	MAXIMUM AISLE escape route
A	30–90 *12–36*	45–75 *18–30*	90 *36*	90–105 *36–42*	120–135 *48–54*
B	35[2] *14*	75 *30*	90 *36*	105 *42*	120 *48*

A PUBLISHED RANGE
B RECOMMENDED
RANGE: Similar sources as used for APP. 1A. Apart from an unnecessary allowance of 90cm/*36″* between-table access to bench seats, no absurdities.
RECOMMENDED: Note that aisle widths accord with general practice but could clearly benefit by being wider, particularly in large establishments.
To an extent, this table repeats information imparted in 5.11, 5.19.

1: Allow 45cm/*18″* for a diner to access a second or further chair at the same table.
2: Increase to 45cm/*18″* if negotiating fixed table/seat units.

CONVERSION TABLES
Length and area

Appendix 2

Feet/in	0	1	2	3	4	5	6	7	8	9	10	11
	Centimeters (cm)											
0	–	2·5	5·1	7·6	10·2	12·7	15·2	17·8	20·3	22·9	25·4	27·9
1	30·5	33·0	35·6	38·1	40·6	43·2	45·7	48·3	50·8	53·3	55·9	58·4
2	61·0	63·5	66·0	68·6	71·1	73·7	76·2	78·7	81·3	83·8	86·4	88·9
3	91·4	94·0	96·5	99·1	101·6	104·1	106·7	109·2	111·8	114·3	116·8	119·4
4	121·9	124·5	127·0	129·5	132·1	134·6	137·2	139·7	142·2	144·8	147·3	149·9
5	152·4	154·9	157·5	160·0	162·6	165·1	167·6	170·2	172·7	175·3	177·8	180·3
6	182·9	185·4	188·0	190·5	193·0	195·6	198·1	200·7	203·2	205·7	208·3	210·8
7	213·4	215·9	218·4	221·0	223·5	226·1	228·6	231·1	233·7	236·2	238·8	241·3
8	243·8	246·4	248·9	251·5	254·0	256·5	259·1	261·6	264·2	266·7	269·2	271·8
9	274·3	276·9	279·4	281·9	284·5	287·0	289·6	292·1	294·6	297·2	299·7	302·3
10	304·8	307·3	309·9	312·4	315·0	317·5	320·0	322·6	325·1	327·7	330·2	332·7
11	335·3	337·8	340·4	342·9	345·4	348·0	350·5	353·1	355·6	358·1	360·7	363·2
12	365·8	368·3	370·8	373·4	375·9	378·5	381·0	383·5	386·1	388·6	391·2	393·7
13	396·2	398·8	401·3	403·9	406·4	408·9	411·5	414·0	416·6	419·1	421·6	424·2
14	426·7	429·3	431·8	434·3	436·9	439·4	442·0	444·5	447·0	449·6	452·1	454·7
15	457·2	459·7	462·3	464·8	467·4	469·9	472·4	475·0	477·5	480·1	482·6	485·1
16	487·7	490·2	492·8	495·3	497·8	500·4	502·9	505·5	508·0	510·5	513·1	515·6
17	518·2	520·7	523·2	525·8	528·3	530·9	533·4	535·9	538·5	541·0	543·6	546·1
18	548·6	551·2	553·7	556·3	558·8	561·3	563·9	566·4	569·0	571·5	574·0	576·6
19	579·1	581·7	584·2	586·7	589·3	591·8	594·4	596·9	599·4	602·0	604·5	607·1
20	609·6	612·1	614·7	617·2	619·8	622·3	624·8	627·4	629·9	632·5	635·0	637·5
21	640·1	642·6	645·2	647·7	650·2	652·8	655·3	657·9	660·4	662·9	665·5	668·0
22	670·6	673·1	675·6	678·2	680·7	683·3	685·8	688·3	690·9	693·4	696·0	698·5
23	701·0	703·6	706·1	708·7	711·2	713·7	716·3	718·8	721·4	723·9	726·4	729·0
24	731·5	734·1	736·6	739·1	741·7	744·2	746·8	749·3	751·8	754·4	756·9	759·5
25	762·0	764·5	767·1	769·6	772·2	774·7	777·2	779·8	782·3	784·9	787·4	789·9
26	792·5	795·0	797·6	800·1	802·6	805·2	807·7	810·3	812·8	815·3	817·9	820·4
27	823·0	825·5	828·0	830·6	833·1	835·7	838·2	840·7	843·3	845·8	848·4	850·9
28	853·4	856·0	858·5	861·1	863·6	866·1	868·7	871·2	873·8	876·3	878·8	881·4
29	883·9	886·5	889·0	891·5	894·1	896·6	899·2	901·7	904·2	906·8	909·3	911·9
30	914·4	916·9	919·5	922·0	924·6	927·1	929·6	932·2	934·7	937·3	939·8	942·3

Square feet	0	1	2	3	4	5	6	7	8	9
	Square metres (m^2)									
0	–	0·09	0·19	0·28	0·37	0·46	0·56	0·65	0·74	0·84
10	0·93	1·02	1·11	1·21	1·30	1·39	1·49	1·58	1·67	1·77
20	1·86	1·95	2·04	2·14	2·23	2·32	2·42	2·51	2·60	2·69
30	2·79	2·88	2·97	3·07	3·16	3·25	3·34	3·44	3·53	3·62
40	3·72	3·81	3·90	3·99	4·09	4·18	4·27	4·37	4·46	4·55
50	4·65	4·74	4·83	4·92	5·02	5·11	5·20	5·30	5·39	5·48
60	5·57	5·67	5·76	5·85	5·95	6·04	6·13	6·22	6·32	6·41
70	6·50	6·60	6·69	6·78	6·87	6·97	7·06	7·15	7·25	7·34
80	7·43	7·53	7·62	7·71	7·80	7·90	7·99	8·08	8·18	8·27
90	8·36	8·45	8·55	8·64	8·73	8·83	8·92	9·01	9·10	9·20
100	9·29	9·38	9·48	9·57	9·66	9·75	9·85	9·94	10·03	10·13

CONVERSIONS FACTORS
length
1 ft = 30.48 cm
1 inch = 2.54 cm
1 cm = 0.39 in
area
1 sq.ft. = 929 cm^2
1 sq.m. = 10.76 ft^2

GEOMETRIC PROPERTIES

Linear ratios: equilateral triangle, square, hexagon, octagon and circle.

Appendix 3

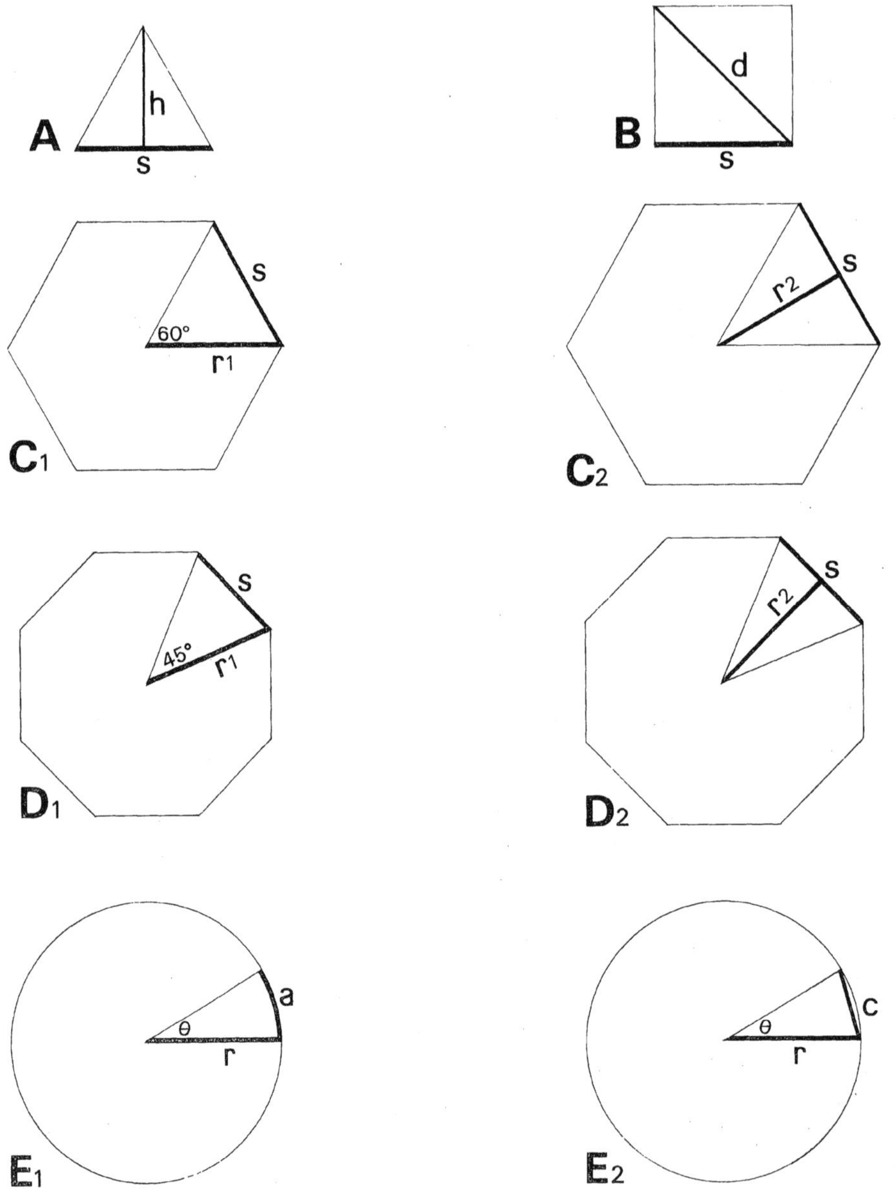

Dimensional ratios and formulae are established by Pythagorean geometry and trigonometry.
To assist visual comparisons, figures are drawn to the same scale: at 1:50, all *given* measurements (drawn extra BOLD) represent a length of 100cm/*40″*.

A **Height** of equilateral triangle.
$h = 0.866\ s$ ($s = 1.155\ h$).

B **Diagonal** of square.
$d = \sqrt{2} = 1.414\ s$.

C1, C2 **Side** of hexagon (derived from equil. triangle).
$s = r^1 = 0.866\ r^2$ ($r^2 = 1.155\ r^1$).

D1, D2 **Side** of octagon (derived from 45° isosceles triangle).
$s = 0.765\ r^1$ and $= 0.828\ r^2$.

E1, E2 **Arc** and **chord** of circle, for certain subtending angles, useful in determining place setting sizes in round tables.

θ°	a	c	relevant to tables seating
30	0.523 r	0.518 r	12
36	0.629 r	0.618 r	10
45	0.785 r	0.766 r	8
60	1.046 r	r	6
90	1.570 r	1.414 r	4

REFERENCES

To aid the reader, major sources are categorised and those found to be of particular use are commented upon.

Space planning (of specific relevance)
*Lawson, F.: *Restaurants, clubs and bars,* London 1987
Tabulations of dining areas, table sizes and planning modules, by a leading British foodservice authority. Much of the information is repeated from the author's previous works and contributions to the New Metric Handbook and to Neufert's Architects Data, and is not infallible.

*Panero, J. (ex Schroeder, F. de N.): *Anatomy for interior designers,* 3rd ed., New York, 1962.
Sets out, in sketch format, the basic human, furniture and space measurements encountered in a series of residential and commercial applications that include dining rooms, restaurants and bars. Though not an in-depth study, and now a bit dated, the authors' perceptive and witty observations – brilliantly accompanied by Nino Repetto's humorous illustrations – will have launched a thousand interior designers on their chosen career. First published 1948, and deservedly still in print after two enlarged and revised editions, this is an unpretentious and delightful masterpiece.

Hepperle, H-A.: *Gaststattenbau,* W. Germany, 1981.
Lawson, F.: *Restaurant planning and design,* London, 1973.
Neufert, E.: *Architects' Data,* (transl.) 1980.
Noble, J.: *Activities and spaces – dimensional data for housing design,* London, 1982.
Ramsey, C. G. and Sleeper, H. R.: *Architectural graphic standards,* 8th ed., New York, 1988.
Reznikoff, S. C.: *Interior graphic and design standards,* New York and London, 1986.
Tutt, P. and Adler, D. (eds.): *New metric handbook,* London, 1979.

Restaurants generally
*Baraban, R. S. and Durocher, J. F.: *Successful restaurant design,* New York, 1989.
Starting from the initial foodservice process, analyses the managerial, strategic, design and contractual issues over a wide márket range. Thorough, thoughtful and informative.
*Cohen, E. L. and Emery, S. R.: *Dining by design,* New York, 1985.
*Colgan, S.: *Restaurant Design,* New York and London, 1987.
*Saito, G. T.: *All about selected American restaurants,* Tokyo, 1986.
Pictorial surveys of contemporary American restaurants with plans, descriptive and analytical texts. Emphasis on style and decor. Published within a year or two of one another, these books are similar in approach and content. Together, they represent an excellent coverage of the more interesting design trends of the mid-eighties.

*Dahinden, J.: *Neue restaurants – ein internationaler querschnitt,* Munich, 1973.
A worldwide survey, comparable to those by Cohen & Emery, Colgan and Saito, but a full decade earlier. An interesting and varied selection of photographs, with an introductory overview (by the well-known Swiss architect author) that succinctly examines the whole gamut of commercial eating facilities.

Gutman, R. J. S. and Kaufman, E.: *American diner,* New York, 1979.
Largely pictorial historical review of the classic American diner, from its late nineteenth century beginnings as a lunch wagon. Superbly nostalgic, a collector's item.

Atkin, W. W. and Adler, J.: *Interiors Book of Restaurants,* New York, 1960.
Fengler, –.: *Restaurant architecture and design,* (transl.) Stuttgart, 1971.
Schirmbeck, E.: *Restaurants – architecture and ambience,* (transl.) Munich, 1982.
Smith, D.: *Hotel and restaurant design,* London Design Council, 1978.

Product manufacturers' literature
*Primo Furniture, U.K.
A major manufacturer and distributor of table-seat combination units and other furniture components for the self-service and fast food sector. A wide range, typically representative of this worldwide specialist industry.

Hostess Furniture, U.K.
Knoll International World Trade.
Wilkhahn, W. Germany.

Anthropometric data
Diffrient, N. et al.: *Humanscale,* Boston, 1974 & 1981.
Dreyfuss, H.: *The Measure of man,* New York, 1971.
Pheasant, S.: *Bodyspace: anthropometry, ergonomics and design,* London, 1988.

MIX
Papier aus verantwortungsvollen Quellen
Paper from responsible sources
FSC® C105338

If you have any concerns about our products,
you can contact us on
ProductSafety@springernature.com

In case Publisher is established outside the EU,
the EU authorized representative is:
Springer Nature Customer Service Center GmbH
Europaplatz 3, 69115 Heidelberg, Germany

Printed by Libri Plureos GmbH
in Hamburg, Germany